Francois Apéry

Models of the Real Projective Plane

Computer Graphics and Mathematical Models

Books

F. Apéry
Models of the Real Projective Plane
Computer Graphics of Steiner and Boy Surfaces

K. H. Becker / M. Dörfler
Computergrafische Experimente mit Pascal.
Ordnung und Chaos in Dynamischen Systemen

G. Fischer (Ed.)
Mathematical Models.
From the Collections of Universities and Museums.
Photograph Volume and Commentary

K. Miyazaki
Polyeder und Kosmos

Plaster Models

Graph of w = 1/z
Weierstraß ℘-Function, Real Part
Steiner's Roman Surface
Clebsch Diagonal Surface

Vieweg

François Apéry

Models of the Real Projective Plane

Computer Graphics of Steiner and Boy Surfaces

With 46 Figures and 64 Color Plates

Friedr. Vieweg & Sohn Braunschweig / Wiesbaden

CIP-Kurztitelaufnahme der Deutschen Bibliothek

Apéry, François:
Models of the real projective plane: computer graphics
of Steiner and Boy surfaces / François Apéry. –
Braunschweig; Wiesbaden: Vieweg, 1987.
  (Computer graphics and mathematical models)

Dr. *François Apéry*
Université de Haute-Alsace
Faculté des Sciences et Techniques
Département de Mathématiques
Mulhouse, France

Vieweg is a subsidiary company of the Bertelsmann Publishing Group.

All rights reserved
© Friedr. Vieweg & Sohn Verlagsgesellschaft mbH, Braunschweig 1987

No part of this publication may be reproduced, stored in a
retrieval system or transmitted in any form or by any means,
electronic, mechanical, photocopying, recording or otherwise,
without prior permission of the copyright holder.

Typesetting: Vieweg, Braunschweig
Printing: Lengericher Handelsdruckerei, Lengerich

ISBN 978-3-528-08955-9   ISBN 978-3-322-89569-1 (eBook)
DOI 10.1007/978-3-322-89569-1

Preface

by *Egbert Brieskorn*

If you feel attracted by the beautiful figure on the cover of this book, your feeling is something you share with all geometers. "It is the delight in figures in a higher sense which distinguishes the geometer". This is a well known statement of the algebraic geometer Alfred Clebsch. Throughout the history of our science, great men like Plato and Kepler and Poincaré were inspired by the beauty and harmony of such figures. Whether we see them as elements or representations of some universal harmony or simply look at them as a structure accessible to our mind by its very harmony, there is no doubt that the contemplation and the creation of beautiful geometric figures has been and will be an integral part of creative mathematical thought.

Perception of beauty is not the work of the intellect alone — it needs sensitivity. In fact, it does need sensual experience. To seperate the ideas from the appearances, the structure from the surface, analytical thinking from holistic perception and science from art does not correspond to the reality of our mind and is simply wrong if it is meant as an absolute distinction. It would deprive our creative thought of one of its deepest sources of inspiration.

I remember a visit of the blind mathematician B. Morin in Göttingen. He gave us a lecture on the eversion of the sphere, the process which he had discovered for turning the sphere inside out by a continuous family of immersions. At that time, he had no analytic description of this very complicated process, but he had a very precise qualitative picture of its geometry in his mind, and he had pictures of models made according to his instruction. I was deeply impressed by the beauty of these figures, and I was moved by the fact that he had been able to see all this beautiful and complex geometry which we could visualize only with great difficulty even after it was shown to us.

In Göttingen we had a very nice collection of mathematical models dating from the days of Felix Klein and David Hilbert. You can find photographs of some of them in a very nice book of Gerd Fischer entitled "Mathematical models", which was published by Vieweg. Among these models there is one made in 1903 showing the surface of Boy. This is a surface obtained by an immersion of the real projective plane in 3-dimensional space. At the time when it was made it was known that the projective plane cannot be embedded as a smooth surface in 3-space. This is so because any smooth closed surface in 3-space divides the space into an interior part and an exterior. This implies that it is orientable: If you stand in the exterior with your feet on the surface, you know what it means to turn left. But the real projective plane is not orientable. This was a very remarkable discovery of Felix Klein in 1874, and it implies that the projective plane $\mathbb{P}^2$ cannot be embedded in the 3-dimensional space $\mathbb{R}^3$. So if we want to visualize $\mathbb{P}^2$ by means of a surface in $\mathbb{R}^3$, we have to be contented with less than a smooth surface. For instance we should admit that different parts of the surface penetrate each other. If the parts themselves are smooth and penetrate each other transversally, we call the surface transversally immersed. It is by no means obvious that there is a presentation of $\mathbb{P}^2$ as an immersed surface in $\mathbb{R}^3$. If there was none, we would be forced to allow the surface to have some singular points, i.e. points where it is not smooth. Such representations of $\mathbb{P}^2$ by surfaces with selfpenetrations and singular points were known to exist. One of them is a beautiful surface discovered by the German geometer Jacob Steiner. He called it his "Roman surface", because he discovered some of its properties during a stay at Tome in 1844. You can see a picture of this surface on **Plate 57** of this book and you can buy a pretty 3-dimensional model of it from Vieweg. It has tetrahedral symmetry, with six lines of selfpenetration issuing from the center of the tetrahedron in the direction of the midpoints. These singular points are of the simplest possible type. They are so called "Whitney-umbrellas" (**Plate 23**). These singularities are the simplest singularities (apart from selfpenetration) which can occur in this situation, and they are stable: a Whitney umbrella will not go away if the surface is deformed slightly.

Maybe it was the stability of these singularities that made David Hilbert suspicious so that he thought it would be impossible to present the projective plane $\mathbb{P}^2$ by an immersed surface without singular points. But the famous mathematician David Hilbert was mistaken. In 1901 his student W. Boy produced two surfaces which are immersions of $\mathbb{P}^2$ in $\mathbb{R}^3$. One of them is particularly pretty. It has a 3-fold axis of symmetry and this is the Boy surface which is the main theme of this book.

Of course, B. Morin knew Boy's surface when he visited us in Göttingen. He must have had a very clear picture of it in his mind. Nevertheless he asked us to take our model out of the show-case so that he could touch it and feel whether its curvature was beautiful. It is this kind of sensitivity for the genesis and transformation of forms which I think is necessary for the creation of ideas such as those of B. Morin or R. Thom leading to the beautiful constructions of F. Apéry that you find in this book.

Boy's surface, like Morin's halfway model of the eversion of the sphere at the time of his visit in Göttingen, was only described in terms of a qualitative geometry. Such descriptions are quite satisfactory from the point of view of differential topology. In fact since the days of H. Poincaré we know that in many situations such a qualitative description is all we can hope for. However, there is an older tradition in mathematics which asks for a more definite way of describing solutions of a problem. In algebraic or analytic geometry, a surface imbedded in 3-space is to be described by an equation or a parametrization by nice analytic functions such as polynomials or by some beautiful geometric construction. In a sense, a geometric figure can only be considered as a unique entity worthy of this name "figure" in the ancient tradition if it is well-defined in this way. This is what Kepler meant when he said: "That which is construction in geometry is consonance in music". Thus our desire for harmony in a geometric figure will only be fulfilled if we see it described in such a way. This beautiful book of F. Apéry tells the exciting history of the discovery of such a description for the surface of Boy.

It begins with very qualitative topological and combinatorical descriptions of the projective plane and the surface of Boy, and it shows us nice descriptions in terms of equations and parametrizations obtained by B. Morin 1978 (**Plate 33**), J.-P. Petit and J. Souriau (1981) (**Plates 31, 32**), J. F. Hughes 1985 (**Plate 34**) and J. Bryant (**Plate 35**). He then presents a very nice solution of his own originating from an idea of B. Morin. The idea is to construct the apparent contour of the Boy surface seen from infinity in the direction of its 3-fold axis by interpreting parts of it as a deformation of the "folded handkerchief singularity" mapping (x, y) to (x^2, y^2). This leads to a parametrization of a Boy surface in 3-space by 3 polynomials of degree 4 (proposition 4 on page 60, **Plates 36–38**). However, all these solutions have slight deficiencies. For instance the equation describing the surface may have other real zeros different from the points of the Boy surface.

But finally the author presents a truly perfect solution. It describes the Boy surface by a beautiful geometric construction, which is subtle and elementary, in the best tradition of classical geometry. In order to make it easy for you to understand his analysis, let me explain the construction in purely geometric terms. It generates the surface by a 1-parameter family of ellipses obtained as follows.

Let L be a line and p a point on L. This line will become the 3-fold axis of symmetry, and the pole p will be a point where L intersects the surface. Let P be the plane orthogonal to L passing through p, and let Q be a plane parallel to P at distance 1. In Q draw a circle K of radius $\sqrt{2}/3$ with axis L. Then draw a threecuspidal hypocycloid C such that K is its inscribed circle. C may be obtained as trace of a point of the circumference of a circle of radius $\sqrt{2}/3$ rolling inside a circle of radius $\sqrt{2}$ concentric with K. The curve C will be the apparent contour of the Boy surface that we are going to construct, if we project the surface from the pole p to the plane Q.

Next we draw an elongated hypocycloid H in the plane Q as follows. Let K' be a circle of radius 1 in Q concentric with K. Let M be a moving circle of radius $1/3$ which rolls without gliding inside K'. Let t_0 be a point in the moving plane of M at distance $\sqrt{2}/3$ from the centre of M. As M rolls, t_0 will trace out an elongated hypocycloid in the plane Q. This curve will have three ordinary double points, and they lie on K. We choose the initial position of t_0 so that these double points are situated on the tangents of the 3 cusps of the 3-cuspidal hypocycloid C, as indicated by the following figure.

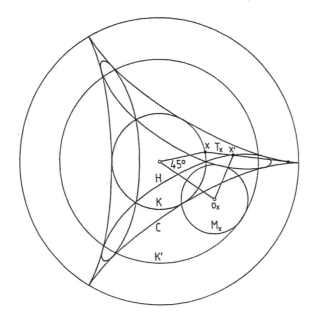

Now we define a map $K \to H$. Let x be a point moving on K in the positive sense. Let o_x be the centre of the moving circle M_x, where o_x is moving on a circle of radius 2/3 with the same angular velocity as x, but trailing behind $45°$. The point t_0 for M_x determines a unique point x' on the elongated hypocycloid H, and our map is defined by associating this point x' to x.

It follows from the construction that x and x' lie on a tangent T_x to the hypocycloid C. Let P_x be the plane passing through this tangent T_x and the pole p. We define an ellipse E_x in the plane P_x by the following three conditions:

(a) E_x passes through x and x'.
(b) E_x is tangent to P in p.
(c) The distance d of x and x' and the maximal distance δ of a point on E_x from P are related as follows:
$$\delta = (1 - 9d^2/16)^{-1}.$$

These conditions define E_x uniquely.

And now our construction comes to its end: The surface S to be constructed is simply the union of all ellipses E_x. As x moves on K, the ellipses sweep out the surface. This is the Boy surface constructed by F. Apéry. Apéry proves that it is exactly the set of real zeroes of a polynomial equation of degree 6 which he can write down explicitly. Moreover he gives a nice description of the transversal immersion $\mathbb{P}^2 \to S \subset \mathbb{R}^3$. The ellipses are the images of a pencil of projective lines passing through the point in $\mathbb{P}^2$ which corresponds to the pole p. If we interpret $\mathbb{P}^2$ as quotient of the 2-sphere S^2 identifying antipodal points, the corresponding map $S^2 \to \mathbb{R}^3$ has a nice description in terms of trigonometric functions, and the inverse images of the ellipses are the great circles passing through a pair of antipodal points. **Plates 39–41** show this pretty surface.

Apéry's construction is beautiful, elementary, but subtle and extremely elegant. Moreover there is a similar construction for the Roman surface of Steiner, and one can pass from one surface to the other by a continuous family of surfaces, such that at a certain moment the six Whitney-umbrella-type singularities on the deformations of the Roman surface annihilate each other in pairs by a nice process called hyperbolic confluence. This process is illustrated on **Plates 28–30,** and the family of surfaces is shown on **Plates 51–56.**

Finally Apéry generalizes his approach and gets similar descriptions of other surfaces. Among them is the beautiful halfway model of B. Morin which you see on the cover of this book. All this is sheer beauty, and I hope that you will be delighted.

Introduction

In this monograph we study the real projective plane from the topological point of view, more precisely we look at its representations in $\mathbb{R}^3$. Provided that we give a sufficiently rich crop of examples, such an excursion into descriptive zoology can serve as the pretext for a didactic introduction to theories with applications extending far beyond the field of the initial investigation.

Guided by this principle we have tried throughout to strike a balance between abstract definitions (from final year undergraduate or first year postgraduate mathematics) and related examples which are enlivened by drawings and computer graphics. The color plates were created using the PS 300 computer of the Laboratoire de Cristallographie de l'Institut de Biologie Moléculaire et Cellulaire in Strasbourg; R. Ripp provided much valuable assistance in taming this remarkable beast.

As soon as it was realized that the real projective plane is nonorientable and can not be embedded in $\mathbb{R}^3$, the hunt was on for tractable representations in three dimensions. The various theoretical stages leading to W. Boy's construction in 1901 of the self-transversal immersion which now bears his name can be conceptualized by the following results of H. Whitney:

(i) any closed surface embeds in $\mathbb{R}^4$. We can obtain an embedding of the projective plane in $\mathbb{R}^4$ from the real Veronese mapping, whose image lies in a 4-dimensional sphere, by stereographically projecting the sphere onto $\mathbb{R}^4$ (sec. 2.3).

(ii) any closed surface embedded or immersed in $\mathbb{R}^4$ can be projected to a locally stable surface in $\mathbb{R}^3$, in other words one whose only singularities are Whitney umbrellas (sec. 2.3). Starting from the embedding described above of the real projective plane in $\mathbb{R}^4$, we find two surfaces originally discovered by Steiner: the Roman surface and the cross-cap (sec. 1.3). Another result of Whitney [WH1] shows that it is impossible to construct an immersion of the projective plane in $\mathbb{R}^3$ by projecting an embedding in $\mathbb{R}^4$.

(iii) Whitney umbrellas occur in pairs on the locally stable image of a closed surface in $\mathbb{R}^3$, and can be eliminated in pairs by means of two generic processes called elliptic and hyperbolic confluence of Whitney umbrellas (sec. 2.3). Starting with the cross-cap, a hyperbolic confluence of the two umbrellas leads to the first of the immersions constructed by Boy **(Plates 115 and 116 [FI])**. If, on the other hand, we start from the Roman surface, we can then retain an axis of threefold symmetry throughout the deformation, and the elimination of the six umbrellas gives the Boy surface with threefold symmetry **(Plates 39–41)**.

We remark that these two constructions in fact lead to the same geometrical object, as we shall show in section 3.2.

We begin chapter 1 by defining surfaces and giving some of their basic properties, the main aim of this chapter is to make an inventory of representations of the real projective plane in $\mathbb{R}^3$, paying particular attention to the Steiner surfaces.

In chapter 2 we explain the concepts of embedded and immersed surfaces in $\mathbb{R}^3$; our goal is to construct a smooth polynomial immersion of $\mathbb{P}^2$ in $\mathbb{R}^3$. We find an immersed surface which is not the real zero-set of any polynomial; there is an irreducible polynomial which vanishes on this surface, but it also vanishes elsewhere in $\mathbb{R}^3$.

As in (iii) above, we then construct a C^1-immersion of $\mathbb{P}^2$ in $\mathbb{R}^3$ by eliminating the six umbrellas from the Steiner Roman surface. The image of this immersion is the real zero-set of a sextic polynomial.

As an interlude, in section 2.1, we give explicit polynomials whose real zero-sets are the closed orientable surfaces in $\mathbb{R}^3$, and also a sextic polynomial whose real zero-set is the image of a smooth self-transversal immersion of the Klein bottle in $\mathbb{R}^3$; this last example is based on an idea of L. Siebenmann.

Chapter 3 is devoted to a classification of certain immersed projective planes in S^3 which are related to the Boy immersion; a final list consists of two examples.

In section 3.3 we obtain the halfway model of an eversion of the sphere described by B. Morin, as a subset of an algebraic surface of degree eight.

We have only introduced abstract mathematical ideas in so far as this was necessary for an understanding of the examples; and we have framed them in such a way that our explanations can be followed by a motivated honours student. We can not hope to give an exhaustive account of this subject and so we refer the interested reader to more technical and more advanced sources.

It was at a showing of a series of slides of surfaces on the occasion of the Symposium on singularities in Warsaw in 1985 that Professor E. Brieskorn offered to put me in touch with the publisher Vieweg in order to produce a book with color plates. In addition to having been at the origin of this book E. Brieskorn was also kind enough to read the manuscript and to allow me to profit from his pertinent comments.

It remains for us to thank A. Flegmann for having rendered this text comprehensible in English.

Table of Contents

Preface (by *Egbert Brieskorn*) V
Introduction IX

**Chapter 1: Some Representations of the Real Projective Plane
before 1900** 1

1.1 Closed Surfaces 1
1.2 The Real Projective Plane 13
1.3 Steiner Surfaces 31

Chapter 2: The Boy Surface 42

2.1 Embedded and Immersed Surfaces 42
2.2 Parametrization of the Boy Surface by Three Polynomials 51
2.3 How to Eliminate Whitney Umbrellas 61
2.4 A Boy Surface of Degree Six 74

Chapter 3: More about Immersions in the 3-dimensional Sphere 82

3.1 Self-Transveral Immersions of the Real Projective Plane 82
3.2 Classification of Immersed Projective Planes of Boy Type 90
3.3 The Halfway Model 103

Appendix: Listing of the FORTRAN Program Used to Draw the Boy Surface
and its Deformations (by *Raymond Ripp*) 107

Bibliography 116
Subject Index 119
Plate Index 121
Plates 123

Chapter 1

Some Representations of the Real Projective Plane before 1900

1.1 Closed Surfaces

In this section our aim is not to give a self-contained treatment of topological manifolds, but to recall some basic facts about curves and surfaces we will use later. The reader will find additional details or proofs in any book of topology or differential topology (see for instance [HI] [GRI] [MAS] [SP]).

A *n-dimensional manifold* M consists of a separable metrizable topological space such that each point admits a neighborhood homeomorphic to the *closed unit ball* $\bar{B}^n$ of $\mathbb{R}^n$.

The *boundary* ∂M of M is defined to be the set of points of M which do not admit any neighborhood homeomorphic to the *open unit ball* B^n. If $\partial M = \phi$, we call M a *manifold without boundary*. The boundary ∂M is closed in M; it is an $(n-1)$-dimensional manifold without boundary, and $M \setminus \partial M$ is an n-dimensional manifold without boundary. The canonical injection $\partial M \to M$ is a *topological embedding*, that is a continuous map which maps ∂M homeomorphically onto its image.

Example 1.

All 0-dimensional manifolds are given by finite or countable discrete topological spaces. These manifolds are without boundary, and compactness is equivalent to finiteness.

A manifold is the *topological sum* of its components [BO], hence the knowledge of connected manifolds is sufficient to describe all manifolds. In this book we shall use mainly 1- or 2-dimensional manifolds, and sometimes 3-dimensional ones.

Example 2.

Every connected 1-dimensional manifold is homeomorphic to one of the following:

> the unit circle S^1 of $\mathbb{R}^2$, compact without boundary
> the line $\mathbb{R}$,   non-compact without boundary
> the half-line $[0, \infty[$, non-compact with boundary
> the *unit segment* $I = [0, 1]$, compact with boundary

A *curve* is a compact 1-dimensional manifold. In the same way, we define a *surface* to be a compact 2-dimensional manifold, and such a manifold is said to be *closed* whenever it is connected without boundary.

Example 3.

The closed unit ball $\bar{B}^n$ of $\mathbb{R}^n$ is a compact connected n-dimensional manifold whose boundary is the closed $(n-1)$-dimensional manifold S^{n-1} called *unit sphere* of $\mathbb{R}^n$.

Example 4.

An *annulus* is a surface homeomorphic to the product $S^1 \times I$. The boundary of an annulus has two components (Fig. 1).

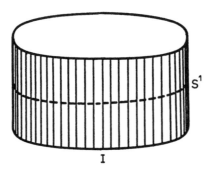

Fig. 1 Annulus

A *disk* is a surface homeomorphic to the closed unit ball $\bar{B}^2$, and an *open disk* is a 2-dimensional manifold homeomorphic to the open unit ball B^2. One way to construct a surface is to glue one elementary surface, for instance a disk, to another one.

To carry out such a gluing, we need a mathematical concept called the *quotient*. If $\sim$ is an equivalence relation on a topological space M, we can characterize the open sets in the *quotient topology* on the quotient set $M/\sim$ as the images of the *saturated* open sets by the *canonical map* $M \xrightarrow{s} M/\sim$. A saturated set in M is a set S which is a union of equivalence classes.

The quotient topology on $M/\sim$ inherits some properties from the topology of M, for instance, connectedness or quasi-compactness, but not Hausdorff's property for which we need this useful proposition:

Proposition 1.

If M is compact, then the quotient topology is Hausdorff if and only if the saturated set generated by any closed set in M is closed. In this case $M/\sim$ is compact.

Example 5.

Let us consider the equivalence relation defined on the square $I \times I$ by $(u, 0) \sim (u, 1)$ and $(0, v) \sim (1, v)$ for every u, v in I. The quotient space $\mathbb{T}^2$ is a closed surface called a *torus*. The map

$$I \times I \xrightarrow{f} \mathbb{R}^3$$

1.1 Closed Surfaces

given by

$$f(u, v) = (\cos 2\pi u \cdot (2 + \cos 2\pi v), \sin 2\pi u \cdot (2 + \cos 2\pi v), \sin 2\pi v)$$

admits a unique factorization through the canonical map from $I \times I$ onto $\mathbb{T}^2$ and a topological embedding of $\mathbb{T}^2$ in $\mathbb{R}^3$ (**Plate 7**).

Example 6.

We consider the equivalence relation on $I \times I$ given by

$$(0, v) \sim (1, 1 - v)$$

for every v in I. The quotient space M, called a *Möbius strip*, is a connected surface whose boundary is S^1. The map

$$f(u, v) = (2 \cos 2\pi u + (2v - 1) \cos \pi u \cos 2\pi u, \ 2 \sin 2\pi u + (2v - 1) \cos \pi u \cdot \sin 2\pi u,$$
$$(2v - 1) \sin \pi u)$$

admits an unique factorization by the canonical map from $I \times I$ onto M and a topological embedding of M into $\mathbb{R}^3$ (Fig. 2).

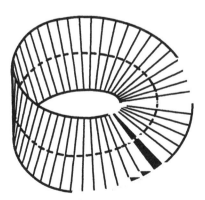

Fig. 2 Möbius strip.

Example 7.

Let D denote the closed unit disk. There are two canonical topological embeddings of D in the topological sum $D + D$ denoted by

$$D \xrightarrow{j} D + D \xleftarrow{k} D.$$

If $S^1 \xrightarrow{i} D$ is the canonical topological embedding, we define an equivalence relation $\sim$ on $D + D$ by $ji(u) \sim ki(u)$ for every u in S^1.

The quotient space $D + D / \sim$ is homeomorphic to the 2-sphere. The sphere is obtained by gluing the northern hemisphere to the southern hemisphere (**Plate 1**).

A *simple loop* on a surface will be a topological embedding of the circle S^1 in the surface. For instance, the map

$$\gamma_1(\theta) = (\theta, 1/2)$$

from S^1 to the annulus of the example 4 defines a simple loop on the annulus. Similarly with the map

$$\gamma_2(\theta) = (2\cos\theta, 2\sin\theta, 0)$$

from the circle to the Möbius strip embedded in $\mathbb{R}^3$ as in the example 6. There is a phenomenon which cannot occur with curves but which is basic in the classification of surfaces, namely orientability. For a surface S we have the key propositions:

Proposition 2.

If γ is a simple loop on S whose image does not meet ∂S, only two disjoint cases can occur:

(i) there is a topological embedding of the annulus in S such that the image of the previous simple loop γ_1 coincides with that of γ.

(ii) there is a topological embedding of the Möbius strip in S such that the image of the previous simple loop γ_2 coincides with that of γ.

Proposition 3.

If γ is a simple loop on S contained in ∂S, then there exists a topological embedding of the annulus in S such that a boundary component coincides with the image of γ.

If there is at least one simple loop on S, as in case (ii) of proposition 2, the surface S is said to be *nonorientable*, otherwise it is *orientable*.

Example 8.

The *Jordan curve theorem* states that if γ is simple loop on the sphere S^2, then there exists a homeomorphism of the sphere into itself which maps the equator to γ. Thus the boundary of a neighborhood of the loop in proposition 2 is not connected, hence we are in case (i). The sphere S^2 is orientable.

In order to achieve the classification of connected surfaces we have to introduce the notions of graph and 2-complex.

1.1 Closed Surfaces

A *graph* is a topological space G with a finite subspace of *vertices* V such that $G \setminus V$ has a finite number of components, called *edges*, and each pair consisting of an edge and its closure is homeomorphic to (I, J) or $(S^1, S^1 \setminus \{1\})$, where J is the interior of the unit segment I. Obviously, a graph is compact.

The closure of an edge is a connected curve, and we can build a graph by gluing curves together in a convenient way. In general, the result doesn't give a curve as shown by the letter "T". On the other hand, a curve can be made into a graph in different ways because it is possible to subdivide any interval with additional vertices.

Example 9.

Let $\sim$ be an equivalence relation on a curve C such that the set of pairs of equivalent points is the union of the diagonal of $C \times C$ and a finite subset. Then the quotient space $C/\sim$ is a graph. For instance, the letter "T" is the quotient of the topological sum of two copies of I by the equivalence which identifies the point 0 in the first factor with 1/2 in the second one (Fig. 3).

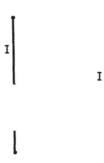

Fig. 3 Letter "T" as a quotient.

Generalizing the definition of a graph, we can characterize a *2-complex* as a topological space K strictly containing a graph G such that $K \setminus G$ has finitely many components, and for each component U there exists a homeomorphism from B^2 onto U which extends continuously to a map from $\bar{B}^2$ to $U \cup G$.

A 2-complex is compact and if a compact space is a 2-complex the 2-complex structure can be realized in infinitely many ways: think about the square $I \times I$ equipped with a tessellation.

There is a nice invariant for the graphs as well as for the 2-complexes, called the *Euler characteristic*. If n_0 is the number of vertices of G, n_1 that of edges, and n_2 that of components of $K \setminus G$, we define the Euler characteristic of K by

$$\chi(K) = n_0 - n_1 + n_2 .$$

If K denotes a graph, we make $n_2 = 0$. It is convenient to call a graph with at least one edge a *1-complex* and a finite set a *0-complex*. By a *complex* we mean any kind of the previous ones.

Proposition 4.

Two complexes of the same homotopy type have the same Euler characteristic.

We say that two topological spaces U and V are of the same *homotopy type* when there are two continuous maps $U \xrightarrow{j} V$ and $V \xrightarrow{k} U$ such that kj is homotopic to the identity map 1_U and jk is homotopic to the identity map 1_V. A *homotopy* between two continuous maps $U \xrightarrow{f_0} V$ and $U \xrightarrow{f_1} V$ is a continuous map $U \times I \xrightarrow{F} V$ such that

$$F(.\,,0) = f_0 \quad \text{and} \quad F(.\,,1) = f_1$$

In this case, f_0 and f_1 are said to be homotopic.

If U and V are homeomorphic, then they are of the same homotopy type, but any point is of the same homotopy type as I, and not homeomorphic to I. We can verify that the Euler characteristics of the point and I are equal to 1.

Example 10.

Let S^1 be the set of complex numbers of absolute value 1. This is a circle on which we can define a graph structure with 1 as unique vertex and with a unique edge. Therefore, the Euler characteristic is 0.

Observe that the Euler characteristic of the topological sum of two circles vanishes as well, although it is not of the same homotopy type as one circle.

Consider the maps $S^1 \xrightarrow{j} A$, $z \mapsto (z, 0)$ where A is the annulus $S^1 \times [-1, 1]$, and $A \xrightarrow{k} S^1$, $(z, x) \mapsto z$. We have $kj = 1_{S^1}$, and

$$F(z, x, t) = (z, tx)$$

gives a homotopy between jk and 1_A. Thus the annulus is of the same homotopy type as the circle, and

$$\chi(A) = 0.$$

In the same way, we consider the maps $S^1 \xrightarrow{j} M$ and $M \xrightarrow{k} S^1$, where M is the image of the Möbius strip by the topological embedding $\bar{f}$ of example 6, and

$$j(e^{2i\pi u}) = f(u, 1/2) \quad kf(u, v) = e^{2i\pi u}$$

(f is defined in example 6). We have $kj = 1_{S^1}$, and an homotopy between jk and 1_M is given by

$$F(f(u, v), t) = f(u, 1/2 + t(v - 1/2))$$

Like the annulus, the Möbius strip is of the same homotopy type as the circle, and

$$\chi(M) = 0.$$

Example 11.

There is a natural 2-complex structure on the disk $\bar{B}^2$ given by $S^1 \setminus \{1\}$. We obtain

$$\chi(\bar{B}^2) = 1 - 1 + 1 = 1.$$

1.1 Closed Surfaces

Example 12.

The north pole determines a 2-complex structure on the sphere S^2 with no edges. It yields

$$\chi(S^2) = 1 - 0 + 1 = 2.$$

Example 13.

We consider the 2-complex structure on the square $I \times I$ given by the four edges. We find

$$\chi(I \times I) = 4 - 4 + 1 = 1.$$

If we make the quotient by the equivalence relation defined in example 5, we obtain a 2-complex structure on the torus $\mathbb{T}^2$ which yields

$$\chi(\mathbb{T}^2) = 1 - 2 + 1 = 0.$$

In fact we have the following proposition:

Proposition 5.

On every surface there exists a 2-complex structure.

Propositions 4 and 5 allow us to define the Euler characteristic of any surface. This leads to the theorem of classification:

Theorem 6.

Two connected surfaces are homeomorphic if and only if they are both orientable or both nonorientable, they have the same Euler characteristic, and they have the same number of boundary components.

One reason to deal with abstract surfaces and not only embedded surfaces in $\mathbb{R}^3$ is the next:

Proposition 7.

A nonorientable closed surface cannot be topologically embedded in $\mathbb{R}^3$.

Sketch of proof.

The space $\mathbb{R}^3$ is separated in two components by the image $f(M)$ of a topological embedding of a closed surface M. For choosing a point m not in the image $f(M)$ we can distinguish the points of the complement of $f(M)$ according to whether they can be joined to m by a path crossing $f(M)$ an even or an odd number of times. So $f(M)$ and consequently M do not contain a Möbius strip. q.e.d.

Example 14.

Let $\sim$ be the equivalence relation on $I \times I$ given by

$$(0, v) \sim (1, 1 - v) \quad \text{and} \quad (u, 0) \sim (1 - u, 1)$$

The quotient space $\mathbb{P}^2$ can be obtained as a quotient of the Möbius strip M of the example 6 by the following equivalence relation on its boundary (identified with the unimodular circle of complex numbers):

$$z \sim \bar{z} \, .$$

By proposition 3 there exists a topological embedding of the annulus into M such that a boundary component of the image A of the annulus coincides with the boundary S^1 of M. We identify A with the annulus of complex numbers given by

$$1 \leqslant |z| \leqslant 2$$

in such a way that S^1 is the unimodular circle. We define the continuous map $A \xrightarrow{f} D$, where D is the disk of center 0 and radius 2, by

$$|x| \geqslant 1 \quad f(z) = z$$
$$|x| \leqslant 1 \quad f(z) = x + i(y - (1-x^2)^{1/2})(4 - x^2 + (1-x^2)^{1/2} \cdot (4-x^2)^{1/2})/3$$

where $x = \operatorname{Re} z$ and $y = \operatorname{Im} z$.

This map factorizes through the canonical map $A \xrightarrow{s} A/\sim$ where $\sim$ denotes the equivalence relation previously defined on S^1. We obtain a surjective continuous map $A/\sim \xrightarrow{\bar{f}} D$:

$$\begin{array}{ccc} A & \xrightarrow{f} & D \\ & \searrow_{s} \quad \nearrow_{\bar{f}} & \\ & A/\sim & \end{array}$$

The equality $f(z_1) = f(z_2)$ is equivalent to $z_1 \sim z_2$, so the map $\bar{f}$ is one-to-one, and proposition 1 proves that $A/\sim$ is compact, so $\bar{f}$ is homeomorphic, and $A/\sim$ is a disk.

Moreover, if we glue a disk along the boundary component S^1 of A, that is if we construct the quotient space of the topological sum $A + \bar{B}^2$ by the equivalence relation which identifies in a trivial way the boundary component S^1 of A with the boundary of $\bar{B}^2$, we obtain again a disk.

We have proved that $\mathbb{P}^2$ is obtained by gluing a disk to a Möbius strip along their boundaries. Considering the equivalence relation on $I \times I$ we have

$$\chi(\mathbb{P}^2) = 2 - 2 + 1 = 1 \, .$$

This surface is nonorientable because it contains a Möbius strip. A surface homeomorphic to $\mathbb{P}^2$ is called a *real projective plane* or just a *projective plane*. In the next section, we shall justify this terminology.

A useful method to build a closed surface is by taking the quotient of the disk, whose boundary is a circle divided into an even number of equal edges, by an equivalence relation which identifies the edges appropriately.

Two continuous maps $[a, b] \xrightarrow{\gamma} X$ and $[c, d] \xrightarrow{\delta} X$ into a topological space X are said to be equivalent whenever there exists an order preserving homeomorphism $[a, b] \xrightarrow{h} [c, d]$ so that $\delta h = \gamma$. An equivalence class is called a *path*. If the previous homeomorphism is order reversing, δ defines the *opposite path* of γ denoted by γ^{-1}. Considering a subdivision of $[a, b]$

$$a = a_0 \leqslant a_1 \leqslant \ldots \leqslant a_{n-1} \leqslant a_n = b$$

1.1 Closed Surfaces

we say that γ is the *sum of the paths* given by the restrictions of γ to each segment $[a_k, a_{k+1}]$. When the restriction of γ to $]a, b[$ is one-to-one, we will say that γ is a *simple path*. When $\gamma(a) = \gamma(b)$, γ defines a *loop* that is a continuous map from the circle into a topological space: if $a < b$, the circle is the quotient of $[a, b]$ by the equivalence relation $a \sim b$.

For instance, consider the unit closed disk $\bar{B}^2$ of complex numbers. It is bounded by the simple loop:

$$\gamma(t) = \exp(2 i \pi t), \qquad t \in [0, 1]$$

We denote by γ_k the restriction of γ to $[(k-1)/2n, k/2n]$ where $1 \leq k \leq 2n$. Then γ is the sum of the simple paths (Fig. 4)

$$\gamma = \gamma_1 \cdots \gamma_{2n}$$

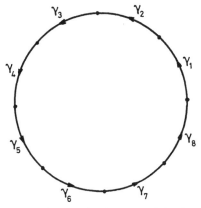
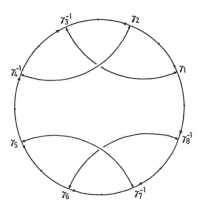

Fig. 4 Disk bounded by a sum of simple paths.

closed surface of characteristic -2 obtained by identifications on the boundary of the disk

Example 15.

$n = 1$.

We define an equivalence relation $\sim$ on $\bar{B}^2$ by $\gamma_1(t) \sim \gamma_2(1-t)$. As we saw in example 14 the quotient space $\bar{B}^2/\sim$ is obtained by gluing a disk to $\bar{B}^2$ along their boundaries. Clearly it yields a sphere, each disk corresponding to a hemisphere and their common boundary to the equator. $\bar{B}^2/\sim$ is a closed orientable surface of characteristic 2.

Example 16.

$n > 0$.

We set

$$\gamma_{2k-1}(t) \sim \gamma_{2k}(t + 1/2n)$$

There are three kinds of points in the quotient space $\bar{B}^2/\sim$. First of all, the points in the open disk. The open disk is a neighborhood of such a point.

Secondly, the equivalence classes (z) such

$$|z| = 1 \text{ and } z^{2n} \neq 1$$

Consider

$$z = \gamma_{2k-1}(t) \quad t \in \,](2k-2)/2n, \,(2k-1)/2n[$$

The class (z) contains exactly z and $z' = \gamma_{2k}(t + 1/2n)$. Let D be the intersection of $\bar{B}^2$ with a small open disk of center z and radius ϵ which contains no $2n-th$ root of unity. We denote by D' the image of D by the rotation sending z to z'. The pairs (D, z) and (D', z') are homeomorphic to $(B^+, 0)$ and $(B^-, 0)$ where B^+ and B^- are the subsets of the open unit disk of complex numbers of imaginary part respectively nonnegative and nonpositive. A neighborhood of (z) is obtained by gluing D to D' along their boundaries, thus it is homeomorphic to the open disk $B^+ \cup B^-$.

And lastly, there is the class (1) containing the $2n-th$ roots of unity. Let D_k be the intersection of $\bar{B}^2$ with a small open disk with center $z_k = \exp(i\pi k/n)$ and radius ϵ such that $D_1, \ldots, D_{2n}$ are disjoint. The pair (D_k, z_k) is homeomorphic to the pair $(B_k, 0)$ where B_k is the sector of the open unit disk defined by

$$z = \rho e^{2i\pi t} \quad 0 \leq \rho < 1 \quad t \in [(k-1)/2n, k/2n]$$

We can choose these homeomorphisms so that the gluing of the D_k's to obtain a neighborhood of (1) corresponds to taking the union of the B_k's, a gluing which yields the open unit disk.

Eventually each point of the quotient space $\bar{B}^2/\sim$ admits a neighborhood homeomorphic to an open disk, and this connected space is compact by proposition 1; thus $\bar{B}^2/\sim$ is a closed surface. The 2-complex structure of $\bar{B}^2$ given by the γ_k's yields

$$\chi(\bar{B}^2/\sim) = 1 - n + 1 = 2 - n$$

Consider A_k the intersection of a small annulus

$$\{z : \epsilon/2 \leq |z - z_{2k-1}| \leq 3\epsilon/2\}$$

with $\bar{B}^2$ such that $A_1, \ldots, A_n$ are disjoint. Each A_k is homeomorphic to the square $I \times I$ so that the identification on the boundary of A_k corresponds to the equivalence relation on the square defined in example 6. So $\bar{B}^2/\sim$ contains n disjoint Möbius strips, and is a nonorientable closed surface of characteristic $2-n$.

Example 17.

$n = 2m > 1$.

We set (Fig. 4)

$$\gamma_{4k-3}(t) \sim \gamma_{4k-1}((8k-5)/4m - t) \qquad \gamma_{4k-2}(t) \sim \gamma_{4k}((8k-3)/4m - t) \quad 1 \leq k \leq m.$$

As previously, there are three kinds of points in the quotient space $\bar{B}^2/\sim$. The verification that each point admits a neighborhood homeomorphic to an open disk is left to the reader. We conclude that $\bar{B}^2/\sim$ is a closed surface of characteristic

$$\chi(\bar{B}^2/\sim) = 1 - 2m + 1 = 2(1-m).$$

1.1 Closed Surfaces

Let $\bar{B}^2 \xrightarrow{f} S$ and $\bar{B}^2 \xrightarrow{f'} S'$ be two topological embeddings of the closed disk into two closed surfaces S and S'. If S^1 denotes the boundary of $\bar{B}^2$, we consider a homeomorphism $fS^1 \xrightarrow{h} f'S^1$ and define $S \# S'$ to be the quotient of the topological sum

$$(S \setminus fB^2) + (S' \setminus f'B^2)$$

by the equivalence relation

$$hf(s) \sim f'(s) \quad s \in S^1$$

Proposition 8.

If $\bar{B}^2 \xrightarrow{g} T$ and $\bar{B}^2 \xrightarrow{g'} T'$ are two topological embeddings in two closed surfaces T and T' homeomorphic to S and S' respectively, then $T \# T'$ is homeomorphic to $S \# S'$. Moreover, $S \# S'$ is a closed surface; it is called the *connected sum* of S and S'. The sum $S \# S'$ is homeomorphic to $S' \# S$, and $S \# (S' \# S'')$ is homeomorphic to $(S \# S') \# S''$, so we can write $S \# S' \# S''$. The connected sum $S \# S'$ is orientable if and only if S and S' are both orientable (**Plate 11**).

Example 18.

The Jordan curve theorem proves that if $\bar{B}^2 \xrightarrow{f} S^2$ is a topological embedding, then $S^2 \setminus fB^2$ is a closed disk. Thus for every closed surface S the connected sum $S \# S^2$ is homeomorphic to S.

Example 19.

We use the notations of example 17 with $m > 1$. Let δ be the path given by

$$\delta(t) = t + (1-t) \exp(i\pi/m) \quad t \in [0, 1].$$

Let M_m be the closed surface $\bar{B}^2/\sim$. We denote by K the compact space bounded by the loop $\gamma_1 \gamma_2 \gamma_3 \gamma_4 \delta$, and by K' the compact subset of the plane bounded by the loop $\gamma'_1 \gamma'_2 \gamma'_3 \gamma'_4 \delta'$ where

$$\gamma'_1(t) = (t, 0) \quad \gamma'_2(t) = (1, t) \quad \gamma'_3(t) = (1-t, 1) \quad \gamma'_4(t) = (0, 1-t)$$
$$\delta'(t) = (1/2) \cos(\pi(1/2 - t)) (\cos(\pi(1-t)/2), \sin(\pi(1-t)/2))$$
$$t \in [0, 1]$$

There exists a continuous map f from K onto K' such that the restriction to $K \setminus \{1, \exp(i\pi/m)\}$ is one-to-one and

$f\gamma_1$ is equivalent to $\gamma'_1, \ldots,$ and $f\delta$ is equivalent to δ'.

We denote by $\sim$ the restriction to K of the equivalence relation defined on $\bar{B}^2$, and also by $\sim$ the restriction to K' of the relation defined on the square in example 5. It is easy to see that there exists exactly one homeomorphism $K/\sim \xrightarrow{\bar{f}} K'/\sim$, such that the following diagram commutes

$$\begin{array}{ccc} K & \xrightarrow{f} & K' \\ {\scriptstyle s}\downarrow & & \downarrow{\scriptstyle s'} \\ K/\sim & \xrightarrow{\bar{f}} & K'/\sim \end{array} \quad \text{where } s \text{ and } s' \text{ are the canonical maps.}$$

Let L denote the compact space bounded by the loop $\gamma_5 \ldots \gamma_{2n} \delta^{-1}$, and $\gamma_1'', \ldots, \gamma_{2n-4}''$ denote the γ_k's where we replace m by $m-1$. We denote by δ'' the loop in the plane given by

$$\delta''(t) = -(1/2)\cos(\pi(1-2t)/2)(\cos(\pi(1-2t)/4), \sin(\pi(1-2t)/4)) \quad t \in [0,1]$$

$L/\sim$ is the quotient space of L by the restriction of the equivalence relation defined on $\bar{B}^2$, and if L'' is the compact space bounded by the loop $\gamma_1'' \ldots \gamma_{2n-4}'' \delta''$, $L''/\sim$ is the quotient of L'' by the equivalence relation defining M_{m-1}. We can build a continuous map g from L to L'' so that the restriction to $L \setminus \{1, \exp(i\pi/m)\}$ is one-to-one and

$$g\gamma_5 \text{ is equivalent to } \gamma_1'', \ldots, g(\delta^{-1}) \text{ is equivalent to } \delta''$$

As before, we see that there exists exactly one homeomorphism $L/\sim \xrightarrow{\bar{g}} L''/\sim$ such that the following diagram is commutative

$$\begin{array}{ccc} L & \xrightarrow{g} & L'' \\ s \downarrow & & \downarrow s'' \\ L/\sim & \xrightarrow{\bar{g}} & L''/\sim \end{array}$$

We now observe that M_m is obtained by gluing $K/\sim$ to $L/\sim$ along the loop $\delta/\sim$, and this is homeomorphic to the gluing of $K'/\sim$ to $L''/\sim$ along the loops δ' and δ''. By definition this gives the connected sum of M_{m-1} and M_1:

$$M_m = M_1 \# M_{m-1} .$$

Using proposition 8, we prove by induction that if M_1 is orientable, then M_m is orientable for every m. Example 5 says that M_1 is a torus, and proposition 7 proves that it is orientable. Eventually the closed surface of example 17 is orientable.

Theorem 9.

Every closed surface is of the type of one of the examples 15, 16, 17. In particular, the Euler characteristic of an orientable closed surface is an even integer less than or equal to 2, and for a nonorientable closed surface it is an integer less than 2.

The next proposition is easy to check with the method developed in example 19.

Proposition 10.

The characteristic of the connected sum of two closed surfaces S and T is given by
$\chi(S \# T) = \chi(S) + \chi(T) - 2$.

Example 20.

We have $\chi(\mathbb{P}^2 \# \mathbb{P}^2 \# \mathbb{P}^2) = -1 = \chi(\mathbb{T}^2 \# \mathbb{P}^2)$ and these two closed surfaces are nonorientable thus $\mathbb{P}^2 \# \mathbb{P}^2 \# \mathbb{P}^2$ is homeomorphic to $\mathbb{T}^2 \# \mathbb{P}^2$.

We can summarize some properties of the Euler characteristic in

1.2 The Real Projective Plane

Proposition 11.

The equivalence classes of closed surfaces up to homeomorphism form a countable commutative semigroup under the operation of connected sum. This semigroup has the sphere as identity element and is generated by the torus $\mathbb{T}^2$ and the projective plane $\mathbb{P}^2$ subject to the single relation

$$\mathbb{P}^2 \# \mathbb{P}^2 \# \mathbb{P}^2 = \mathbb{T}^2 \# \mathbb{P}^2$$

Proposition 11 shows that each closed surface can be written in a unique way as a connected sum of tori or as a connected sum of projective planes. The number of terms in such a decomposition is called the *genus* of the closed surface. In particular, the genus of the sphere yields 0, that of the torus and the projective plane yields 1 (**Plates 1, 7, 10, 12**).

The nonorientable closed surface of genus 2 is called the *Klein bottle*. The properties of the projective plane and the connected sum show that the Klein bottle is obtained by gluing two Möbius strips along their boundaries (**Plates 14–16**).

1.2 The Real Projective Plane

Just as the adjunction of imaginary numbers has simplified the study of algebraic equations, so the creation of the projective geometry by Gérard Desargues in the beginning of the 17th century, by adjunction of elements at infinity to an affine plane, has simplified certain problems of intersection by cancelling the concept of parallelism.

Though the use of the projective geometry goes back a long time, the first appearance of the real projective plane as a surface, in the sense of the previous section, seems to have been due to August Ferdinand Möbius.

In a work entitled „Mémoire sur les polyèdres" proposed to the Académie des Sciences de Paris for the Grand Prix de Mathématiques in 1861, Möbius constructs a surface by gluing five triangles, in order to obtain the so-called Möbius strip (Fig. 5), and then considers the cone over this strip. Then Möbius proves that the surface so built is closed nonorientable and of characteristic 1 [PO]. Unfortunately, he did not recognize the projective plane.

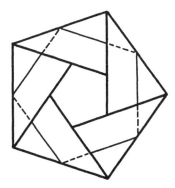

Fig. 5 Möbius strip obtained by gluing five triangles.

We have to credit Felix Klein with the discovery in 1874 of the projective plane as a closed nonorientable surface of characteristic one [KL].

There are many ways to represent the projective plane as a closed surface. Of course all of them give the same surface in the sense of sect. 1.1, but each is of interest in view of specific topological properties. That is why we are going to describe some of them.

Representation 1 (Möbius).

This representation consists of gluing a disk to a Möbius strip along their boundaries. This method is described in Fig. 6. Ex. 14 sect. 1.1 proves we obtain the projective plane $\mathbb{P}^2$.

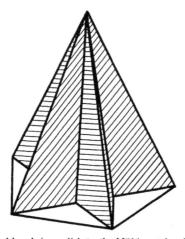

Fig. 6 Projective plane obtained by gluing a disk to the Möbius strip of Fig. 5 along their boundaries.

Representation 2 (Klein).

We begin to build a 2-complex structure on the projective plane starting from the real affine plane p and by adding elements at infinity following the method of Desargues.

To each affine line l on p we join a point at infinity denoted by $i(l)$ so that $i(l) = i(l')$ if and only if l is parallel to l'. We define a *projective line*, or just a *line*, to be either the union of an affine line and its point at infinity, or the set

$$\lambda = \{i(l): l \text{ affine line}\}.$$

A *homography* is a one-to-one map from $p \cup \lambda$ onto itself, which transforms any line into a line; and the group of the homographies is called the *projective group*. A *flag* on $p \cup \lambda$ is given by a pair (μ, α) where μ is a line and α a point on μ. The projective group acts transitively on the set of flags, that is: for every flag (μ', α') there exists a homography h so that $h\mu = \mu'$ and $h\alpha = \alpha'$.

1.2 The Real Projective Plane

Let (x, y) be a coordinate system on the affine plane. The map

$$k(x, y) = (1 + x^2 + y^2)^{-1/2} \cdot (x, y)$$

is onto-to-one from p onto the disk B^2. The image of an affine line l passing through the origin is a diameter of B^2 bounded by two antipodal points on the boundary S^1 denoted by $a(l)$ and $b(l)$.

The inverse map k^{-1} is a homeomorphism from B^2 onto p. We extend k^{-1} to the closed disk $\bar{B}^2$ by

$$k^{-1}(a(l)) = k^{-1}(b(l)) = i(l) .$$

We provide $p \cup \lambda$ with the finest topology making k^{-1} to be continuous.

This topology induces the canonical topology on p. It makes the homographies continuous, and hence they are homeomorphisms. Furthermore, $\bar{B}^2$ being connected and compact and k being surjective, $p \cup \lambda$ is connected and compact. The map k proves that each point on p admits a neighborhood homeomorphic to a disk, and, by the transitive action of the projective group, $p \cup \lambda$ is a closed surface.

By the transitive action of the projective group on the flags we see that a line is a 1-dimensional manifold homeomorphic to a fixed line of the type $l \cup i(l)$ where l is an affine line. We note the line remains connected when we remove the point $i(l)$, it is thus homeomorphic to the circle S^1 (ex. 2 sect. 1.1).

It follows

$$\chi(p \cup \lambda) = 1 - 1 + 1 = 1 .$$

By sect. 1.1 $p \cup \lambda$ is homeomorphic to $\mathbb{P}^2$. That is why we call $\mathbb{P}^2$ the projective plane.

Representation 3 (Walther von Dyck [DY]).

The method of von Dyck consists of taking the quotient of the closed disk $\bar{B}^2$ by the antipodal equivalence relation on its boundary S^1. We can consider a homeomorphism from $\bar{B}^2$ onto the square $I \times I$ such that the antipodal relation on the boundary S^1 gives the equivalence relation on $I \times I$ given in the beginning of ex. 14 sect. 1.1. This example shows that the quotient of the closed disk by the antipodal relation on its boundary is the projective plane $\mathbb{P}^2$.

Using the continuous map k^{-1} defined in representation 2, we have the following factorization

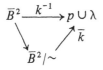

by the quotient surjection and a bijection $\bar{k}$, where $\sim$ is the antipodal relation on S^1. In fact, $\bar{k}$ is a homeomorphism for which a line passing through the origin corresponds to a diameter of $\bar{B}^2$ on which the extremities have been identified.

Representation 4 (Hermann Grassmann [GRA]).

The idea of Grassmann lies in considering the set G of the 1-dimensional vector subspaces of $\mathbb{R}^3$. To give a topology on G we express it as the quotient of $\mathbb{R}^3 \setminus \{0\}$ by the equivalence relation $\sim$ for which the equivalence classes are the elements of G:

$$(x, y, z) \sim (x', y', z') \quad \text{when} \quad x/x' = y/y' = z/z'$$

according to the convention such that a term in a quotient vanishes if and only if the other one in the same quotient vanishes also. We can provide G with the quotient topology on $(\mathbb{R}^3 \setminus \{0\})/\sim$. As in Möbius's case Grassmann did not see that his construction gives the projective plane. To show this we use the map f from $\mathbb{R}^3 \setminus \{0\}$ to the space $p \cup \lambda$ of representation 2, so that

$$f(x, y, z) = (x/z, y/z) \quad \text{when} \quad z \neq 0$$
$$f(x, y, 0) = i(l) \quad \text{when} \quad z = 0$$

where l denotes any affine line whose direction is given by the vector (x, y). The map f factorizes through the canonical surjection from $\mathbb{R}^3 \setminus \{0\}$ to G giving a bijection $\bar{f}$ from G to $p \cup \lambda$:

$$\begin{array}{ccc}
\mathbb{R}^3 \setminus \{0\} & \xrightarrow{f} & p \cup \lambda \\
& \searrow \quad \nearrow \bar{f} & \\
& G &
\end{array}$$

By the bijection $\bar{f}$, a 2-dimensional vector subspace of $\mathbb{R}^3$ corresponds to a line of $p \cup \lambda$, and a linear automorphism of $\mathbb{R}^3$ corresponds to a homography. Using the fact that $\bar{f}$ is locally open and continuous at the class of $(0, 0, 1)$, and the transitivity of the actions of the linear group on $\mathbb{R}^3 \setminus \{0\}$ and the projective group on $p \cup \lambda$, we prove $\bar{f}$ is a homeomorphism. So G is homeomorphic to the projective plane $\mathbb{P}^2$.

Representation 5.

Let $\sim$ the antipodal relation on the sphere S^2. By prop. 1 sect. 1.1 the quotient space $S^2/\sim$ is compact. Furthermore, the restriction of the canonical surjection from S^2 onto $S^2/\sim$ to every open hemisphere is a homeomorphism onto its image, so $S^2/\sim$ is a closed surface.

We represent S^2 as the unit sphere in $\mathbb{R}^3$, and we denote by i the canonical injection from S^2 to $\mathbb{R}^3 \setminus \{0\}$. By the universal property of the quotient space we have the following commutative diagram

$$\begin{array}{ccc}
S^2 & \xrightarrow{i} & \mathbb{R}^3 \setminus \{0\} \\
\downarrow & & \downarrow \\
S^2/\sim & \xrightarrow{j} & G
\end{array}$$

where G is the space of representation 4, the vertical arrows are the quotient surjections and j is a continuous bijection.

1.2 The Real Projective Plane

The compactness of $S^2/\sim$ implies j is a homeomorphism, so $S^2/\sim$ is a representation of the projective plane $\mathbb{P}^2$.

Using representation 4 we see that by j, a line corresponds to the quotient of a great circle of S^2 by the antipodal relation.

Representation 6 (Curt Reinhardt [RE]).

We start with the graph given by the edges and vertices of a regular octahedron, and we construct a 2-complex structure by adding four triangular faces of the octahedron, chosen so that any two have at most one common vertex, and three square faces corresponding to the diagonal of the octahedron (Fig. 7).

It is easy to prove we obtain a closed surface H, and we have

$$\chi(H) = 6 - 12 + 7 = 1$$

The 2-complex H is called the *Rheinhardt heptahedron*. It is homeomorphic to $\mathbb{P}^2$.

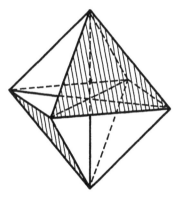

Fig. 7 Rheinhardt heptahedron.

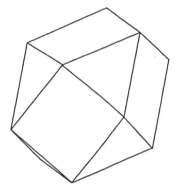

This heptahedron can be described as the quotient of the cuboctahedron by the antipodal map.

Representation 7.

As in the preceding representation, we build a 2-complex whose vertices are those of a regular octahedron 0. We specify two antipodal vertices as the north and the south pole, so the four remaining vertices define the equatorial square. To the twelve edges of the octahedron, we add the two diagonals of the equatorial square. To complete the structure we add the four faces of 0 containing the south pole, two faces of 0 with just the north pole as common vertex, two new faces defined by the north pole and the diagonals of the equatorial square, and a quadrilateral face bounded by the two equatorial edges still free and the two previous diagonals (Fig. 8).

As previously, this 2-complex, called the *combinatorial cross-cap* and denoted by C, is a closed surface and its Euler characteristic yields

$$\chi(C) = 6 - 14 + 9 = 1.$$

Thus C is homeomorphic to $\mathbb{P}^2$.

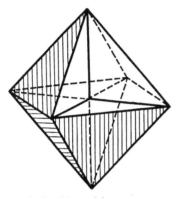

Fig. 8 Combinatorial cross-cap.

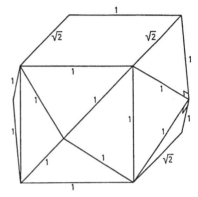

The combinatorial cross-cap can be obtained by making the quotient of this 2-complex by the antipodal map. We have replaced four lateral faces of a rectangular parallelepipet by four pyramids.

Representation 8.

We consider the equivalence relation on the *complex projective line* $\mathbb{C} \cup \{\infty\}$ given by

$$z \sim -1/\bar{z} \quad \text{and} \quad 0 \sim \infty$$

$\mathbb{C} \cup \{\infty\}$ is the Alexandroff compactification of $\mathbb{C}$; so prop. 1 sect. 1.1 implies the compactness of the quotient $\mathbb{C} \cup \{\infty\}/\sim$. This leads back to the spherical model of representation 5 by the stereographic projection from S^2 to $\mathbb{C} \cup \{\infty\}$ (Fig. 9):

$$p(x, y, z) = (x + iy)(1 - z)^{-1} \quad \text{and} \quad p(0, 0, 1) = \infty$$

This homeomorphism induces a homeomorphism between $(\mathbb{C} \cup \{\infty\})/\sim$ and the quotient of S^2 by the antipodal relation.

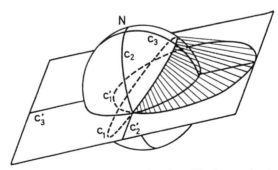

Fig. 9 Stereographic projection from the sphere to the plane. The image of a circle c is denoted by c'.

1.2 The Real Projective Plane

The image of a great circle on S^2 by p is either an affine line containing the origin 0 and completed by ∞, or a circle with center z and radius $1 + z\bar{z}$, according to the fact that the great circle goes or not through the north pole $(0, 0, 1)$ (Fig. 9). It gives the image of a line in this representation.

Representation 9.

Let K be the 2-complex, in Fig. 10. We consider the equivalence relation $\sim$ on K which identifies the edges according to the numbering and the orientation indicated on the figure. The particularity of this 2-complex is the possibility of realizing the former identifications by folding paper according to instructions given in Fig. 10.

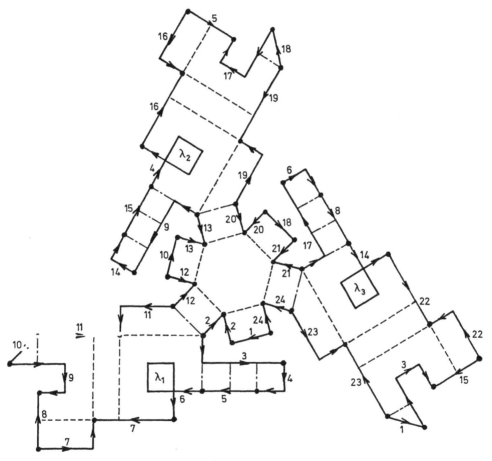

Fig. 10 The numbering of the edges of this 2-complex defines an equivalence relation which can be realized by folding the paper along the dashed lines. The two types of dashed lines indicate the direction of the folding: $----$ for $\vee$ folding $-\circ-$ for $\wedge$ folding. The self-intersection curve in the paper model is the image of the loop defined by the sum $(6-) \cdot \lambda_1 \cdot (6+) \cdot (8) \cdot (14) \cdot (15) \cdot (4-) \cdot \lambda_2 \cdot (4+) \cdot (5) \cdot (6) \cdot (8) \cdot (14-) \cdot \lambda_3 \cdot (14+) \cdot (15) \cdot (4) \cdot (5)$.
where
$$(6-) \cdot (6+) = (6) \qquad (4-) \cdot (4+) = (4) \qquad (14-) \cdot (14+) = (14)$$

It is easy to see that the quotient space $K/\sim$ is homeomorphic to the quotient of the disk $\bar{B}^2$ by the antipodal relation on the boundary (Fig. 11), so $K/\sim$ is homeomorphic to the projective plane $\mathbb{P}^2$ (see representation 3). By prop. 7 sect. 1.1 there are some points of self-intersection in the paper model in $\mathbb{R}^3$. In Fig. 10 these points come from the loop given by the sum

$$\lambda = (6-) \cdot \lambda_1 \cdot (6+) \cdot (8) \cdot (14) \cdot (15) \cdot (4-) \cdot \lambda_2 \cdot (4+) \cdot (5) \cdot (6) \cdot (8) \cdot \\ \cdot (14-) \cdot \lambda_3 \cdot (14+) \cdot (15) \cdot (4) \cdot (5)$$

where

$$(6-) \cdot (6+) = (6) \\ (4-) \cdot (4+) = (4) \\ (14-) \cdot (14+) = (14)$$

and (i) denotes the path whose number is i.

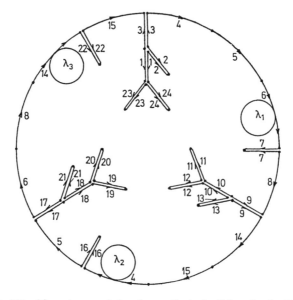

Fig. 11 The model of Fig. 10 can be regarded as the quotient of a disk under the identifications shown.

The image of $K/\sim$ in $\mathbb{R}^3$ obtained by folding paper is a polyhedron called the *combinatorial Boy surface* (Fig. 12). The image of λ, that is the self-intersection curve, defines a loop in $\mathbb{R}^3$ which is called a *three-bladed propeller* (Fig. 13). It has only one multiple point, which is a triple point.

1.2 The Real Projective Plane

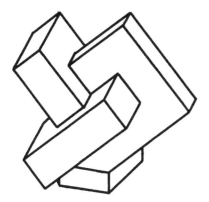

Fig. 12 Combinatorial model of the Boy surface constructed in Fig. 10.

Fig. 13 Self-intersection curve of the model of Fig. 12: this is a three-bladed propeller.

Representation 10.

Let K be the 2-complex of Fig. 14. The complex K is isomorphic to the 2-complex K' of Fig. 15 (by an isomorphism of 2-complex we mean a homeomorphism which restricts to a homeomorphism on the underlying graphs and a bijection on the sets of vertices). The isomorphism is fixed by the numbering and the orientation of the edges on each figure.

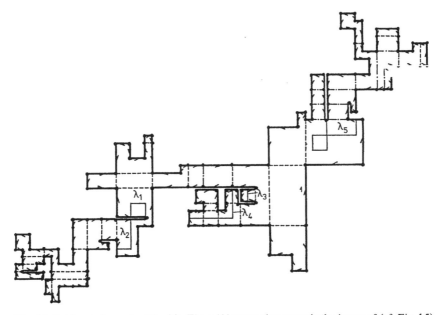

Fig. 14 Fold and glue as for Fig. 10. The self-intersection curve is the image of (cf. Fig. 15)
(50) . (13) . (21) . (22) . (23) . (24) . (25) . (26) . (27) . (28) . (29) .
(50) . (13) . (21) . (22) . (23) . (24) . (25) . (26) . (27) . (28) . (29) .
(33) . (45) . (47) . (49) . $\lambda_1 . \lambda_2 . \lambda_3 . \lambda_4 . \lambda_5 . (32)^{-1} . (31)^{-1} . (30)^{-1}$

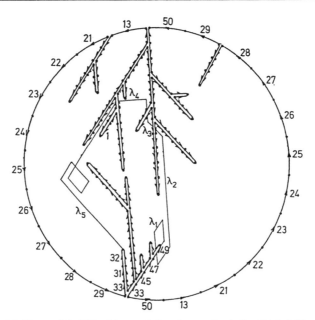

Fig. 15 The model of Fig. 14 regarded as the quotient of a disk (cf. Fig. 11).

We define the equivalence relation $\sim$ on K' which identifies the edges according to numbering and orientation. The quotient space $K'/\sim$ is homeomorphic to the projective plane $\mathbb{P}^2$ as in representation 3. As in representation 9, we can realize the identifications on the 2-complex K with respect to the equivalence relation induced by $\sim$ and the isomorphism from K to K', by folding paper following the instructions of Fig. 14.
In this case, the set of self-intersection points comes from the loop

$$(\lambda) = (50) . (13) . (21) . (22) . (23) . (24) . (25) . (26) . (27) . (28) . (29) .$$
$$(50) . (13) . (21) . (22) . (23) . (24) . (25) . (26) . (27) . (28) . (29) .$$
$$(33) . (45) . (47) . (49) . \lambda_1 . \lambda_2 . \lambda_3 . \lambda_4 . \lambda_5 . (32)^{-1} . (31)^{-1} . (30)^{-1}$$

where (i) denotes the path whose number is i. The image of K' in $\mathbb{R}^3$ so obtained is a polyhedron P whose self-intersection curve is again a three-bladed propeller (Fig. 16).

It is time to define the concept of fundamental group of a *pathwise connected* pointed space (X, x_0). A topological space is pathwise connected if any two points can be joined by a path; and if we fix a *base-point* x_0, we say the space is pointed. On such a space we define a *loop based at* x_0 to be a continuous map λ from the unit circle S^1 of $\mathbb{C}$ in X so that $\lambda(1) = x_0$.

Two such loops λ_1 and λ_2 are said to be *equivalent* whenever there exists a homotopy F between λ_1 and λ_2 preserving the basepoint, that is

$$F(1, t) = x_0 \quad \text{for every } t \text{ in } I.$$

1.2 The Real Projective Plane 23

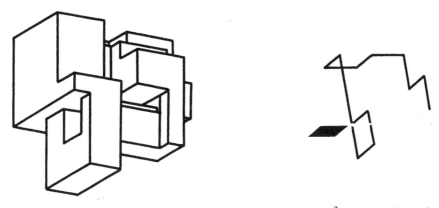

Fig. 16 Nonstandard combinatorial model of the projective plane in $\mathbb{R}^3$ corresponding to the construction of Fig. 14, and its self-intersection curve.

The sum of two loops based at x_0, is defined as in sec. 1.1, as is the idea of the opposite loop. This operation makes the set of equivalent classes of loops based at x_0 to be a group [MAS], where the inverse is given by the opposite loop and identity is the class of the *constant loop*

$$\lambda(z) = x_0 \quad \text{for every } z \text{ in } S^1$$

This group is called the *fundamental group* of (X, x_0), denoted by $\pi(X, x_0)$; we shall write $[\lambda]$ for the *class of* λ.

Example 1.

Let X be a *starlike* subset of $\mathbb{R}^n$ from a point x_0, that is an union of segments $[x_0, x]$. If λ is a loop of X based at x_0, we can consider the homotopy $S^1 \times I \xrightarrow{F} X$

$$F(z, t) = (1-t)\lambda(z) + t \cdot x_0 .$$

This homotopy proves that the fundamental group is trivial

$$\pi(X, x_0) = 1 .$$

If we consider a path $I \xrightarrow{\gamma} X$ so that

$$\gamma(0) = x_0 \quad \gamma(1) = x_1$$

we can define the map

$$\pi(X, x_0) \xrightarrow{\gamma_*} \pi(X, x_1)$$

by

$$\gamma^*[\lambda] = [\gamma^{-1}\lambda\gamma] .$$

Proposition 1.

The map γ^* is an isomorphism of groups.

The proof is trivial. This proposition shows that the fundamental group is independent of the base-point.

Proposition 2.

Two pathwise connected spaces of the same homotopy type have isomorphic fundamental groups.

Proof.

We consider two pathwise connected spaces X, Y and two continuous maps $X \xrightarrow{j} Y$ and $Y \xrightarrow{k} X$ such there exist homotopies $X \times I \xrightarrow{J} X$ and $Y \times I \xrightarrow{K} Y$ such that

$$J(\,.\,,0) = kj \qquad J(\,.\,,1) = 1_X$$
$$K(\,.\,,0) = jk \qquad K(\,.\,,1) = 1_Y$$

Let x_0 be a base-point on X. We write $y_0 = jx_0$, $x_1 = ky_0$, $y_1 = jx_1$. Let γ and δ be the paths in X and Y given by

$$\gamma(t) = J(x_0, t) \quad \delta(t) = K(y_0, t)\,.$$

We define the maps $\pi(Y, y_0) \xrightarrow{f} \pi(X, x_1)$, $\pi(X, x_0) \xrightarrow{g} \pi(Y, y_0)$, $\pi(X, x_1) \xrightarrow{h} \pi(Y, y_1)$ by

$$f[\lambda] = [k(\lambda)] \quad g[\lambda] = [j(\lambda)] \quad h[\lambda] = [j(\lambda)]\,.$$

The maps f, g, h are morphisms of groups, and we have the following commutative diagram

$$\begin{array}{ccc} \pi(X, x_0) & \xrightarrow{\gamma^*} & \pi(X, x_1) \\ {\scriptstyle g}\downarrow & {\scriptstyle f}\nearrow & \downarrow {\scriptstyle h} \\ \pi(Y, y_0) & \xrightarrow{\delta^*} & \pi(Y, y_1) \end{array}$$

From the previous proposition we know that γ^* and δ^* are isomorphisms of groups, so f is a group isomorphism. q.e.d.

Example 2.

By example 1, an open or closed disk has a trivial fundamental group. By the previous proposition, every topological space homeomorphic to an open or closed disk has a trivial fundamental group.

1.2 The Real Projective Plane

Example 3 (The fundamental group of the circle).

Consider a loop $I \xrightarrow{\lambda} S^1$ with $\lambda(0) = \lambda(1) = 1$. It is a property of the exponential function that if α is a continuous function there exists exactly one factorization

$$\begin{array}{ccc} I & \xrightarrow{\lambda} & S^1 \\ {}_\alpha \searrow & & \nearrow_e \\ & \mathbb{R} & \end{array}$$

such that $\alpha(0) = 0$ and $e(t) = \exp(2i\pi t)$. We have $e(\alpha(1)) = \lambda(1) = 1$, so $\alpha(1)$ is an integer called the *degree* of λ and denoted by deg λ. If the family $\{\lambda_t\}_{t \in I}$ realizes a homotopy from $\lambda = \lambda_0$ to the loop λ_1 in S^1 which preserves the base-point 1, then deg λ_t depends continuously on t.

The group $\mathbb{Z}$ of integers being discrete, deg λ_t is constant. Thus we can define the degree deg $[\lambda]$ of a class $[\lambda]$. We see that the degree defines a morphism of groups from $\pi(S^1, 1)$ to $\mathbb{Z}$. The surjectivity of this morphism is proved by the formula

$$\deg[\lambda] = n \quad \text{where} \quad \lambda(t) = \exp(2i\pi nt).$$

Let λ be a loop in S^1 based at 1 so that deg $[\lambda] = 0$. We write

$$\lambda(t) = e(\alpha(t)),$$

and we define a homotopy between λ and the constant loop preserving the base-point 1 by

$$F(t, u) = e((1-u)\alpha(t)) \qquad u \in I.$$

Thus $[\lambda] = [1]$, and we have proved that the degree is an isomorphism of groups from the fundamental group of S^1 onto the infinite cyclic group:

$$\pi(S^1, 1) = \mathbb{Z}.$$

Proposition 2 shows that the fundamental group of an annulus is also $\mathbb{Z}$.

Proposition 3.

Let U and V be two pathwise connected open subsets of the topological space X such that $X = U \cup V$, that the intersection $U \cap V$ is nonempty and pathwise connected, and such that the fundamental groups of U and V are trivial. Then X is pathwise connected and its fundamental group is trivial.

Proof.

Let $S^1 \xrightarrow{\lambda} X$ be a loop based at x_0 in U. The compactness of S^1 allows the decomposition of λ as a sum of paths

$$\lambda = \delta_0 . \delta_0' \ldots \delta_{p-1}' . \delta_p$$

such that the *supports* (that is the images) of $\delta_0, \ldots, \delta_p$ are contained in U, and those of $\delta_0', \ldots, \delta_{p-1}'$ in V.

Because $U \cap V$ is pathwise connected there exists a path γ_k in $U \cap V$ having the same origin and extremity as δ_k'. The loop λ is homotopic to

$$\delta_0 \gamma_0 \gamma_0^{-1} \delta_0' \ldots \gamma_{p-1} \gamma_{p-1}^{-1} \delta_{p-1}' \delta_p$$

with respect to the base-point x_0 (Fig. 17), furthermore the triviality of the fundamental group of V implies that the loop $\gamma_k^{-1} \delta_k'$ is homotopic to the constant loop with respect to the extremity of δ_k'. Then λ is homotopic to $\delta_0 \gamma_0 \ldots \gamma_{p-1} \delta_p$ with respect to x_0, and the triviality of the fundamental group of U implies that $\delta_0 \gamma_0 \ldots \gamma_{p-1} \delta_p$ is homotopic to the constant loop with respect to x_0, and so we have proved that the fundamental group of X is trivial. q.e.d.

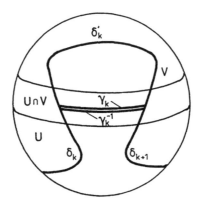

Fig. 17 U and V are two pathwise connected open subsets which cover the sphere. Each has trivial fundamental group and their intersection is nonempty and pathwise connected; this shows that the sphere has trivial fundamental group.

Example 4.

Let N and S be the complements of the poles in the n-dimensional sphere S^n. The sets N and S are open and verify the conditions of proposition 3 for $n \geq 2$, so

$$\pi(S^n, x_0) = 1 \text{ for } n \geq 2.$$

One way of computing fundamental groups is the use of coverings.

Let $X \xrightarrow{p} B$ be a continuous surjective map from the Hausdorff space X onto the pathwise connected Hausdorff space B. We shall say that X is a *covering of B of projection p and base B*, whenever every point b in B admits an open pathwise connected neighborhood U such that p restricts to an homeomorphism on each component of $p^{-1}(U)$. If b belongs to B, we call the subset $p^{-1}b$ of X the *fiber above b*.

Proposition 4.

Every fiber of a covering is homeomorphic to a fixed discrete topological space which we shall call the *fiber of the covering*.

1.2 The Real Projective Plane

Proof.

The fiber above a point b_0 is *discrete*, that is each point in the fiber has a neighborhood which meets no other point of the fiber. Furthermore, the subset of points b such that the fiber above b is homeomorphic to that above b_0 is nonempty, open and closed in the connected space B. This subset must therefore be the whole of B. q.e.d.

Example 5.

Let F be a discrete topological space and B be a pathwise connected Hausdorff space. The product space $B \times F$ provided with the canonical projection $B \times F \xrightarrow{p} B$ defines a covering of base B fiber F and projection p, called a *trivial covering*.

Example 6.

The exponential map $\mathbb{R} \xrightarrow{e} S^1$ used in example 3 defines a covering of base S^1, projection e, and fiber $\mathbb{Z}$.

When the fiber of a covering is a finite set of n elements, we speak of an *n-sheeted covering*.

Example 7.

Consider the notations of example 6 sect. 1.1. The continuous map f can be extended to $[0,2] \times I$ and verifies

$$f(u+1, v) = f(u, 1-v) \quad \text{for} \quad 0 \leq u \leq 1.$$

Then the image remains unchanged, and

$$f(2, v) = f(0, v),$$

so we have the following factorization

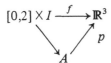

where A denotes the annulus, quotient of $[0,2] \times I$ by the identification

$$(0, v) \sim (2, v).$$

We see that p is a 2-sheeted covering of the Möbius strip by the annulus.

Example 8.

In representation 5, the projective plane is the quotient of the sphere S^2 by the antipodal relation, and the quotient surjection gives a 2-sheeted covering of $\mathbb{P}^2$ by S^2.

Example 9.

The map f of example 5 sect. 1.1 extends to $\mathbb{R}^2$ and gives an infinite sheeted covering of the torus $\mathbb{T}^2$ embedded in $\mathbb{R}^3$ by $\mathbb{R}^2$, whose fiber above $(3, 0, 0)$ is the subgroup $\mathbb{Z}^2$ of $\mathbb{R}^2$.

Proposition 5 (homotopy lifting property).

Let $X \xrightarrow{p} B$ be a covering, $x_0 \in X$, $b_1 \in B$, and $b_0 = p x_0$. Let $I \times I \xrightarrow{F} B$ be a homotopy preserving extremities, that is

$$F(0, t) = b_0 \quad F(1, t) = b_1 \quad \text{for every } t.$$

Then there exists exactly one homotopy $I \times I \xrightarrow{G} X$ so that $pG = F$ and $G(0, t) = x_0$ for every t.

In addition, $G(1, t)$ does not depend on t.

Proof.

The compactness of $I \times I$ allows the existence of two subdivisions of I

$$0 = s_0 \leqslant \ldots \leqslant s_m = 1$$
$$0 = t_0 \leqslant \ldots \leqslant t_n = 1$$

such that for every i and j the image $F([s_{i-1}, s_i] \times [t_{j-1}, t_j])$ is contained in an open subset $U_{i,j}$ of B for which p restricts to a homeomorphism on each component of $p^{-1}(U_{i,j})$. By hypothesis we have $G(s_0, t_0) = x_0$. We define $V_{i,j}$ as the component of $p^{-1}(U_{i,j})$ containing $G(s_{i-1}, t_{j-1})$ assumed well determined. Thus we write

$$G(s, t) = p^{-1} \circ F(s, t) \quad (s, t) \in [s_{i-1}, s_i] \times [t_{j-1}, t_j].$$

By induction on i and j, we construct the desired homotopy G. On each tile $[s_{i-1}, s_i] \times [t_{j-1}, t_j]$, G is uniquely determined by induction, this completes the proof.   q.e.d

We use the notations of proposition 5. If λ is a loop on B, based at b_0, and x a point of the fiber F above b_0, the previous proposition shows that there exists exactly one path γ in X of origin x such that $\lambda = p\gamma$ (consider the constant homotopy $\Phi(s, t) = \lambda(s)$).

Furthermore we know that the extremity of γ depends only on the class $[\lambda]$ in $\pi(B, b_0)$, so we denote it by $x * [\lambda]$. We have

$$x * [b_0] = x \quad (x * [\lambda]) * [\mu] = x * ([\lambda] \cdot [\mu]),$$

in particular

$$(x * [\lambda]) * [\lambda]^{-1} = x = (x * [\lambda]^{-1}) * [\lambda],$$

thus the map $[\tilde{\lambda}]: x \mapsto x * [\lambda]$ is a *permutation* of the fiber F, that is a bijection from F onto F. The map $[\lambda] \mapsto [\tilde{\lambda}]$ is a antimorphism from the fundamental group $\pi(B, b_0)$ to the group of permutations $\mathfrak{S}(F)$:

$$([\lambda][\mu])^{\sim} = [\tilde{\mu}] \circ [\tilde{\lambda}] \quad [\tilde{b}_0] = 1_F.$$

We define the morphism of groups $\pi(X, x_0) \xrightarrow{p\#} \pi(B, b_0)$ by $p\#[\mu] = [p\mu]$, and the map $\pi(B, b_0) \xrightarrow{\partial} F$ by $\partial[\lambda] = x_0 * [\lambda]$ where F is the fiber above b_0. We write $\text{Ker}(p\#) = \{[\mu]: [p\mu] = [b_0]\}$ and $\text{Ker} \, \partial = \{[\lambda]: x_0 * [\lambda] = x_0\}$. When X is connected we speak of a *connected covering*.

1.2 The Real Projective Plane

Proposition 6.

If $X \xrightarrow{p} B$ is a connected covering and $px_0 = b_0$ then the sequence
$1 \to \pi(X, x_0) \xrightarrow{p\#} \pi(B, b_0) \xrightarrow{\partial} (F, x_0) \to 1$ is *exact* that is $\operatorname{Ker}(p\#) = 1$,
$\operatorname{Im}(p\#) = \operatorname{Ker} \partial$, and $\operatorname{Im} \partial = F$.

Proof.

First of all we note that X is connected and locally pathwise connected, so it is pathwise connected. Let x be a point in the fiber above b_0, and let γ be a path joining x_0 to x. We define the loop $\lambda = p\gamma$ on B based at b_0. We have $x = x_0 * [\lambda]$, so $\operatorname{Im} \partial = F$.
The equality $p\#[\mu] = [b_0]$ is equivalent to $[p\mu] = [b_0]$, and by proposition 5 it is equivalent to $[\mu] = [x_0]$, so $\operatorname{Ker}(p\#) = 1$. Clearly $\operatorname{Im}(p\#) \subseteq \operatorname{Ker} \partial$, and the inverse inclusion comes from the definition of ∂. q.e.d.

For instance, consider example 3. We have a connected covering with $\mathbb{R} \xrightarrow{e} S^1$ and $\pi(\mathbb{R}, 0) = 1$. Thus $\pi(S^1, 1) \xrightarrow{\partial} (F, 0)$ is a bijection, and we see that the group structure on F making ∂ a morphism is the canonical group structure on $\mathbb{Z}$. So

$$\pi(S^1, 1) = \mathbb{Z}.$$

In the same way, the covering of example 9 gives

$$\pi(\mathbb{T}^2, x_0) = \mathbb{Z}^2.$$

In example 8 we have an 2-sheeted connected covering of $\mathbb{P}^2$ by S^2, and example 4 proves that $\pi(S^2, x_0) = 1$. Thus $\pi(\mathbb{P}^2, b_0)$ has two elements, that is

$$\pi(\mathbb{P}^2, b_0) = \mathbb{Z}/2.$$

Example 10.

The exponential map $\mathbb{C} \xrightarrow{\exp} \mathbb{C} \setminus \{0\}$ furnishes the projection of an infinite sheeted connected covering of $\mathbb{R}^2 \setminus \{0\}$ by $\mathbb{R}^2$.

Proposition 7.

S^2 and $\mathbb{P}^2$ can be covered by S^2. All other closed surfaces can be covered by $\mathbb{R}^2$.

Proof.

We shall proceed by induction using the decomposition of a closed surface as a connected sum of tori or projective planes according to prop. 11 sect. 1.1. Previous examples prove that the statement is true for S^2, $\mathbb{P}^2$ and $\mathbb{T}^2$.

We now assume that $T_1 \xrightarrow{p_1} S_1$ and $T_2 \xrightarrow{p_2} S_2$ are two coverings of two closed surfaces by S^2 or $\mathbb{R}^2$. In order to compute the connected sum $S_1 \# S_2$ we consider two disks in S_1 and S_2 whose boundaries are the supports of two simple loops λ_1 and λ_2. The fundamental group of a disk being trivial we have $[\lambda_1] = [s_1]$ $[\lambda_2] = [s_2]$ where s_1, s_2 are base-points in S_1 and S_2.
Using prop. 5 we construct two simple loops μ_1, μ_2 on T_1 and T_2 so that $p_1 \mu_1 = \lambda_1$, $p_2 \mu_2 = \lambda_2$.

By Jordan curve theorem on S^2 or $\mathbb{R}^2$ there exist two disks E_1 and E_2 in T_1 and T_2 whose boundaries are the supports of μ_1 and μ_2. The restrictions q_1 and q_2 of p_1 and p_2 to E_1 and E_2 are projections of coverings. In addition, a covering being a local homeomorphism, the inverse images of the supports of λ_1 and λ_2 by q_1 and q_2 are contained in those of μ_1 and μ_2. So $E_1 \xrightarrow{q_1} q_1 E_1 = D_1$ and $E_2 \xrightarrow{q_2} q_2 E_2 = D_2$ are one-sheeted coverings.

The compactness of E_1 and E_2 implies that q_1 and q_2 are homeomorphisms and D_1 and D_2 are disks whose boundaries coincide with the supports of λ_1 and λ_2. Then we see that the connected sum $T_1 \# T_2$ built with E_1 and E_2 is a covering of $S_1 \# S_2$ built with D_1 and D_2. We have

$S^2 \# S^2$ homeomorphic to S^2
$S^2 \# \mathbb{R}^2$ homeomorphic to $\mathbb{R}^2$
$\mathbb{R}^2 \# \mathbb{R}^2$ homeomorphic to $\mathbb{R}^2 \setminus \{0\}$

Example 10 shows that there exists a covering of $\mathbb{R}^2 \setminus \{0\}$ by $\mathbb{R}^2$. This completes the proof by induction.   q.e.d.

Corollary 8.

A simple loop on a closed surface is homotopic to the constant loop if and only if its support is the boundary of a disk.

Proof.

Let λ be a simple loop on the surface S whose support is the boundary of a disk D. We saw in example 2 that the fundamental group of D is trivial, so λ is homotopic to the constant loop.

For the converse we use a covering of S by S^2 or $\mathbb{R}^2$, and the result is contained in the proof of prop. 7.   q.e.d.

In the case of the projective plane we know that there exist exactly two classes of loops in the fundamental group. The class of the constant loop is characterized by the loops whose supports are boundaries of disks. Such a loop shall be called an *oval*.

Proposition 9.

A simple loop on $\mathbb{P}^2$ is an oval if and only if it admits a neighborhood homeomorphic to an annulus.

Proof.

According to prop. 2 sect. 1.1, assume that the simple loop λ admits a neighborhood homeomorphic to a Möbius strip. Then the complement of the support of λ is a connected open subset whose adherence is $\mathbb{P}^2$. So λ does not bound any disk and is not an oval.

Conversely, let λ be a simple loop which admits a neighborhood homeomorphic to an annulus A. Using the definition of the connected sum and the fact that A has two boundary components, we see that $\mathbb{P}^2$ is the connected sum of two closed surfaces along λ.

1.3 Steiner Surfaces

By prop. 11 sect. 1.1, one of these closed surfaces is the sphere S^2. The support of λ is the common boundary of two disks in S^2 and the boundary of one of them in $\mathbb{P}^2$.

q.e.d.

Consider the annulus A on the sphere S^2 bounded by two lines of latitude parallel and equidistant from the equator. The quotient of A by the antipodal relation is a Möbius strip, as we can see from example 7. Representation 5 shows that the image of the equator by the covering projection from S^2 onto $\mathbb{P}^2$ is a projective line. Thus a projective line on $\mathbb{P}^2$ admits a neighborhood homeomorphic to a Möbius strip. So a simple loop on $\mathbb{P}^2$ which is not not homotopic to the constant loop shall be called a *line*.

1.3 Steiner Surfaces

In the previous section, we gave some topological representations of the projective plane as a closed surface. In this section, we want to describe the projective plane as a part of the zero set of a polynomial in $\mathbb{R}^3$. By prop. 7 sect. 1.1 such a description is not one-to-one. Before 1900, such representations were mainly given by the surfaces discovered by the geometer Jacob Steiner in the middle of the nineteenth century. Let us give the historical origin.

To construct these surfaces, Steiner started from a property proved in 1816 by M. Frégier [FR]:

Let (q) be a non-degenerate quadric, 0 be a point on (q), (c) be a conic contained in a plane (π) which does not meet 0, and let (τ) be a tetrahedron inscribed in (q), having 0 as a vertex, whose three edges issuing from 0 meet (π) along a self-polar triangle with respect to (c). Then the side of (τ) opposite to 0 passes through a fixed point F independent of the chosen tetrahedron (Fig. 18).

This property was already a generalization of an analogous property in the projective plane, but the lack of appreciation of matrix calculus at this time did not allow Frégier

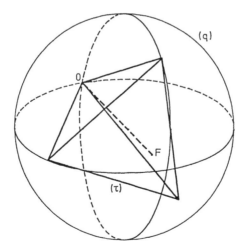

Fig. 18 The tetrahedron (τ) inscribed in the quadric (q) has a fixed vertex at 0. The three edges meeting at 0 are constrained to intersect a fixed plane in a triangle which is self-polar with respect to a fixed conic. The face of the tetrahedron opposite 0 passes through the fixed point F: this is the Frégier point.

to extend it to higher dimensions. This is the general form we can give for an algebraically closed field of characteristic different from 2: (the *n-dimensional projective space* is defined as in rep. 4 sec. 1.2).

Proposition 1.

Let (q) be a non-degenerate quadric hypersurface in the n-dimensional projective space, 0 be a point on (q), (c) be a non-degenerate hyperconic (quadric of codimension 2) contained in a hyperplane (π) which does not meet 0. Let (τ) be a n-simplex inscribed in (q), whose n edges issuing from its vertex 0 meet (π) along a self-polar $(n-1)$-simplex with respect to (c). Then the hyperplane side of (τ) opposite to 0 passes through a fixed point F, independent of the chosen (τ) and called the *Frégier point*.

Proof.

Let $M = (u_1 | \ldots | u_n)$ be an $n \times n$-matrix whose columns $u_1, \ldots, u_n$ are homogeneous coordinates of the vertices of a self-polar $(n-1)$-simplex with respect to (c). If C denotes the matrix of (c), we have $\det C \neq 0$, and we may arrange, by multiplying the homogeneous coordinates of the $(n-1)$-simplex's vertices by a non-zero constant, that:

$${}^tMCM = I \cdot \det C \quad \text{where } I \text{ is the unity } n \times n\text{-matrix.}$$

Permute the columns of M if necessary to make $\det M = 1$; now we know

$$M\widetilde{M} = \widetilde{M}M = I \cdot \det M$$

where $\widetilde{M}$ is the transpose of the matrix of cofactors of M. It follows that

$${}^t\widetilde{M} \cdot \det C = CM.$$

We write the matrix of (q)

$$\left(\begin{array}{c|c} Q & \begin{matrix} v^1 \\ \vdots \\ v^n \end{matrix} \\ \hline v^1 \cdots v^n & 0 \end{array} \right).$$

If $\begin{pmatrix} u_j^1 \\ \vdots \\ u_j^n \end{pmatrix}$ is the jth column of M, then we can write the equations

of the corresponding edge issuing from 0 of the n-simplex obtained by adjoining 0 to the previous $(n-1)$-simplex

$$\frac{x^1}{u_j^1} = \cdots = \frac{x^n}{u_j^n}$$

1.3 Steiner Surfaces

If we denote by 2λ the common value of these quotients, we see that this edge meets (q) in a point $(2\lambda u_j^1, \ldots, 2\lambda u_j^n, x^{n+1})$ so that

$$0 = (2\lambda u_j^1, \ldots, 2\lambda u_j^n, x^{n+1}) \begin{pmatrix} & & v^1 \\ & Q & \vdots \\ & & v^n \\ \hline v^1 \cdots v^n & 0 \end{pmatrix} \begin{pmatrix} 2\lambda u_j^1 \\ \vdots \\ 2\lambda u_j^n \\ x^{n+1} \end{pmatrix}$$

$$= 4\lambda \left[\lambda (u_j^1, \ldots, u_j^n) Q \begin{pmatrix} u_j^1 \\ \vdots \\ u_j^n \end{pmatrix} + x^{n+1} \langle u_j, v \rangle \right]$$

where

$$\langle u_j, v \rangle = (u_j^1, \ldots, u_j^n) \begin{pmatrix} v^1 \\ \vdots \\ v^n \end{pmatrix} .$$

λ being different from zero, we find the point F_j with the homogeneous coordinates

$$(2 u_j^1 \langle u_j, v \rangle, \ldots, 2 u_j^n \langle u_j, v \rangle, - {}^t u_j Q u_j) ;$$

the quantity $\langle u_j, v \rangle$ is not equal to zero whenever we require that the previous edge is not contained in the tangent hyperplane to (q) in 0. Let $(u_0^1, \ldots, u_0^n)$ be the pole of the intersection of (π) with this tangent hyperplane with respect to (c). We may assume

$$C \begin{pmatrix} u_0^1 \\ \vdots \\ u_0^n \end{pmatrix} = \det C \begin{pmatrix} v^1 \\ \vdots \\ v^n \end{pmatrix}$$

that is

$$\begin{pmatrix} u_0^1 \\ \vdots \\ u_0^n \end{pmatrix} = \widetilde{C} \begin{pmatrix} v^1 \\ \vdots \\ v^n \end{pmatrix}$$

The equations of the line joining 0 to this pole are

$$\frac{x^1}{u_0^1} = \cdots = \frac{x^n}{u_0^n} ,$$

which give the parametric representation $x^1 = \lambda u_0^1, \ldots, x^n = \lambda u_0^n$.
We want to prove that the point of intersection F of this line with the hyperplane determined by $F_1, \ldots, F_n$ does not depend on the coefficients u_j^i of M. The coordinates of F are $(\lambda u_0^1, \ldots, \lambda u_0^n, x^{n+1})$, and verify

$$0 = \begin{vmatrix} 2 u_1^1 \langle u_1, v \rangle & \cdots & 2 u_1^n \langle u_1, v \rangle - {}^t u_1 Q u_1 \\ 2 u_n^1 \langle u_n, v \rangle & \cdots & 2 u_n^n \langle u_n, v \rangle - {}^t u_n Q u_n \\ \lambda u_0^1 & \cdots & \lambda u_0^n & x^{n+1} \end{vmatrix} = \Delta .$$

We have det $M = 1$, and we expand Δ by the last line

$$\Delta = 2^n \langle u_1, v\rangle \ldots \langle u_n, v\rangle x^{n+1} + 2^n \lambda \sum_{j=1}^{n} (-1)^{n-j} u_0^j \Delta^j$$

where

$$\Delta^j = \begin{vmatrix} u_1^1 \langle u_1, v\rangle & \ldots & u_1^{j-1} \langle u_1, v\rangle & u_1^{j+1} \langle u_1, v\rangle & \ldots & u_1^n \langle u_1, v\rangle & {}^t u_1 Q u_1 \\ u_n^1 \langle u_n, v\rangle & \ldots & u_n^{j-1} \langle u_n, v\rangle & u_n^{j+1} \langle u_n, v\rangle & \ldots & u_n^n \langle u_n, v\rangle & {}^t u_n Q u_n \end{vmatrix}.$$

The expansion of Δ^j by the last column yields

$$\Delta^j = \sum_{i=1}^{n} (-1)^{n-i} \, {}^t u_i Q u_i \, \Delta_i^j \langle u_1, v\rangle \ldots \langle u_{i-1}, v\rangle \langle u_{i+1}, v\rangle \ldots \langle u_n, v\rangle$$

where Δ_i^j is the minor of the term of the ith row and the jth column of ${}^t M$, that is, from the relation $\widetilde{M} \det C = {}^t M C$

$$\Delta_i^j = \det^{-1} C \, ({}^t M C)_i^j \, (-1)^{i+j}$$

where M_i^j denotes the term of the ith row and the jth column of M. Then

$$\Delta = 2^n \langle u_1, v\rangle \ldots \langle u_n, v\rangle x^{n+1}$$

$$+ 2^n \lambda \det^{-1} C \sum_{i,j} u_0^j \langle u_1, v\rangle \ldots \langle u_{i-1}, v\rangle \langle u_{i+1}, v\rangle \ldots \langle u_n, v\rangle \, {}^t u_i Q u_i \, ({}^t M C)_i^j$$

Now

$$\sum_j u_0^j ({}^t M C)_i^j = ({}^t M C)_i \, u_0 = ({}^t M C)_i \, \widetilde{C} v = ({}^t M C \widetilde{C})_i v = \det C \, ({}^t M v)_i = \det C \langle u_i, v\rangle$$

Thus

$$\Delta = 2^n \langle u_1, v\rangle \ldots \langle u_n, v\rangle \left(x^{n+1} + \lambda \sum_i {}^t u_i Q u_i \right)$$

We infer from the relation $\Delta = 0$

$$x^{n+1} = -\lambda \sum_i {}^t u_i Q u_i = -\lambda \sum_i ({}^t M)_i Q (M)^i = -\lambda \, tr \, ({}^t M Q M) = -\lambda \, tr \, (Q M \, {}^t M)$$

Now

$$C \, {}^t M = \det C \cdot {}^t \widetilde{M} \, {}^t M = \det C \cdot I$$

Hence

$$M \, {}^t M = \widetilde{C},$$

and the coordinates of F are given by

$$\begin{pmatrix} x^1 \\ \vdots \\ x^n \end{pmatrix} = \widetilde{C} \begin{pmatrix} v^1 \\ \vdots \\ v^n \end{pmatrix} \quad \text{and} \quad x^{n+1} = -tr \, (Q\widetilde{C}).$$

q.e.d.

1.3 Steiner Surfaces

Remark 2.

The Frégier point lies on the line joining 0 to the pole of the intersection (d) of (π) with the tangent hyperplane to (q) in 0, with respect to (c). If, in the system of homogeneous coordinates $(x^1, \ldots, x^{n+1})$, the point 0 is $(0, \ldots, 0, 1)$ and the equation of the hyperplane (π) is $x^{n+1} = 0$, if C denotes the matrix of (c) with respect to $(x^1, \ldots, x^n)$ and Q that of the intersection of (q) with (π), and if the equation of the tangent hyperplane to (q) in 0 is $v_1 x^1 + \ldots + v_n x^n = 0$, then the homogeneous coordinates of the Frégier point are given by

$$\begin{pmatrix} x^1 \\ \vdots \\ x^n \end{pmatrix} = \tilde{C} \begin{pmatrix} v^1 \\ \vdots \\ v^n \end{pmatrix} \qquad x^{n+1} = -tr(Q\tilde{C})$$

where trM denotes the trace of M.

We note that these coordinates depend linearly on the coefficients of (q) and are homogeneous polynomials of degree $n - 1$ in the coordinates of (c).

Furthermore, the condition for the hyperconic (c) to be nondegenerate can be improved in view of the coordinates of the Frégier point. Indeed, if we denote by (d) the intersection, of (π) with the tangent hyperplane to (q) in 0, and by (γ) that of (q) and (d) ((γ) is a 3-codimensional quadric), then we have

Proposition 3.

The Frégier point is defined if and only if the hyperconic (c) has at most one singular point, and if this point does not lie on (γ).

Proof.

First of all, consider the case where (q) is not tangent to (π). We can choose the projective frame satisfying the conditions of proposition 1 such that

$$Q = I \quad \text{and} \quad \begin{pmatrix} v_1 \\ \vdots \\ v_n \end{pmatrix} = \begin{pmatrix} 0 \\ \vdots \\ 0 \\ 1 \end{pmatrix}.$$

The Frégier point is not defined if and only if

$$tr\tilde{C} = 0 \quad \text{and} \quad \tilde{C} \begin{pmatrix} 0 \\ \vdots \\ 0 \\ 1 \end{pmatrix} = \begin{pmatrix} 0 \\ \vdots \\ 0 \\ 0 \end{pmatrix}.$$

The Cramer formulae show that the second condition is equivalent to (c) admitting a singular point on the hyperplane $x^n = 0$, that is on the intersection (d) of this hyperplane with (π).

By the same Cramer formulae, the first condition ($tr\widetilde{C} = 0$) is equivalent to (c) and $(\widetilde{c})$ admitting a common singular point, where $(\widetilde{c})$ is the tangential hyperconic defined as the set of points whose polars with respect to

$$(x^1)^2 + \ldots + (x^n)^2 = 0$$

are tangent to (c). That means, in the case where (c) contains singular points, the polars pass through these points.

It follows that $tr\widetilde{C} = 0$ is equivalent to the fact that (c) admits a singular point on $(q) \cap (\pi)$ whose equations are precisely

$$x^{n+1} = 0 \quad \text{and} \quad (x^1)^2 + \ldots + (x^n)^2 = 0$$

The Frégier point is not defined if and only if either (c) admits an unique singular point, this one lying on $(\gamma) = (q) \cap (d)$, or (c) admits more than one singular point (in which case $\widetilde{C} = 0$).

In the case where (q) is tangent to (π), then, (q) being non-degenerate, we have $rkQ = n-1$. Furthermore (π) not passing through 0, the tangent hyperplane to (q) in 0 does not pass through the contact point of (q) with (π), and we can choose projective coordinates such that

$$Q = \begin{pmatrix} 1 & & & 0 \\ & \ddots & & \\ & & 1 & \\ 0 & & & 0 \end{pmatrix} \quad \text{and} \quad \begin{pmatrix} v^1 \\ \vdots \\ v^n \end{pmatrix} = \begin{pmatrix} 0 \\ \vdots \\ 0 \\ 1 \end{pmatrix}$$

The Frégier point is not defined if and only if

$$tr Q\widetilde{C} = 0 \quad \text{and} \quad \widetilde{C} \begin{pmatrix} 0 \\ \vdots \\ 0 \\ 1 \end{pmatrix} = \begin{pmatrix} 0 \\ \vdots \\ 0 \end{pmatrix}$$

which gives a matrix $\widetilde{C}$ of the form

$$\widetilde{C} = \begin{pmatrix} & & & 0 \\ & & & \vdots \\ & & & 0 \\ \hline 0 & \cdots & 0 & 0 \end{pmatrix}$$

The first condition then gives $tr\widetilde{C} = 0$. We therefore have the same conclusion as in the first case.   q.e.d.

Equipped with this result of Frégier in dimension 3, Steiner studied the surface traced out by the Frégier point as the conic (c) describes a *2-pencil of conics*, that is a linear family with two parameters generated by three independent conics. In such a 2-pencil there are degenerate conics, and proposition 3 allows us to choose it in such a way so that the Frégier point is always defined. It suffices to ensure that the 2-pencil does not contain any double line, and that neither of the two points of the 3-codimensional quadric (γ) is a singular point of some conic in the 2-pencil. We have just built a *Steiner surface*.

1.3 Steiner Surfaces

When he visited Rome in 1844, Steiner discovered some geometrical properties of the surface he had constructed according to the above principles. So he knew for instance that it was generated by conics. Indeed, if the conic (c) describes a *1-pencil*, that is a one-parameter linear family generated by two independent conics, then by remark 2 the Frégier point depends quadratically on this parameter and describes a conic.

It seems that Steiner never wrote any article about this surface that he called his *Roman surface* (Plates 21, 22). Since he was not very familiar with complicated computations, he couldn't have determined either its degree in full assurance or any parametric representation. A year before his death, Steiner asked K. Weierstrass to carry on these computations; Weierstrass was able to do so without any problems, knowing that the coordinates of the Frégier point are quadratic forms in the coefficients of the conic (c), as we saw in remark 2 [WE].

The surface thus obtained is a rational algebraic surface whose parametrization is everywhere defined; it is thus of degree four. The computations of Weierstrass suggest that we define a *Steiner surface* to be an algebraic surface of 3-dimensional complex projective space parametrized by four independent quadratic forms which do not vanish simultaneously. This amounts to projecting the *Veronese surface* V, parametrized in the 5-dimensional complex projective space by $x^2, y^2, z^2, yz, zx, xy$, to a 3-dimensional one by a projection whose center is a line not meeting V.

The simplicity of this definition allows to reach easily the characteristic properties of the Steiner surfaces. First of all, these are of degree four. The parametrization of a Steiner surface defines a *3-pencil of conics* with no fixed point in the complex projective space. The classification of the 3-pencils of conics with no fixed point up to homography is a preliminary step towards the classification of Steiner surfaces.

Lemma 4.

A 3-pencil of conics contains at least one double line.

Proof.

Let f_1, f_2, f_3, f_4 be four independent quadratic forms which generate the 3-pencil of conics. The pencil necessarily contains a degenerate conic. Assume this is not a double line and set

$$f_1 = 2xy$$
$$f_i = a_i x^2 + b_i y^2 + 2c_i xz + 2d_i yz + e_i z^2 \qquad i = 2, 3, 4$$

If the determinant

$$D = \begin{vmatrix} c_1 & d_1 & e_1 \\ c_2 & d_2 & e_2 \\ c_3 & d_3 & e_3 \end{vmatrix}$$

vanishes, then the pencil contains a conic defined by $ax^2 + by^2$, and the 1-pencil $[2xy, ax^2 + by^2]$ contains a double line. If $D \neq 0$, we may suppose that

$$f_2 = a_2 x^2 + b_2 y^2 + 2xz$$
$$f_3 = a_3 x^2 + b_3 y^2 + 2yz$$
$$f_4 = a_4 x^2 + b_4 y^2 + z^2 \ .$$

The conic $f_4 + \lambda_3 f_3 + \lambda_2 f_2 + 2\lambda_1 xy$ is a double line if and only if

$$\lambda_1 = \lambda_2 \lambda_3 \qquad \lambda_2 = \lambda_2 a_2 + \lambda_3 a_3 + a_4 \qquad \lambda_3^2 = \lambda_2 b_2 + \lambda_3 b_3 + b_4 \ .$$

This system admits at least one solution in $\lambda_1, \lambda_2, \lambda_3$. Thus, the pencil contains a double line. q.e.d.

Lemma 5.

A 3-pencil of conics with no fixed point contains at least two double lines.

Proof.

The previous lemma allows us to set

$$f_1 = x^2$$
$$f_i = 2 a_i xy + b_i y^2 + 2 c_i xz + 2 d_i yz + e_i z^2$$

for $i = 2, 3, 4$. If the determinant

$$D = \begin{vmatrix} c_2 & d_2 & e_2 \\ c_3 & d_3 & e_3 \\ c_4 & d_4 & e_4 \end{vmatrix}$$

vanishes, then the pencil contains a conic of the type $2axy + by^2$. If $b \neq 0$, the 1-pencil $[x^2, 2axy + by^2]$ contains two double lines. If $b = 0$, we can take $f_2 = 2xy$, $a_3 = a_4 = 0$. If the determinant

$$E = \begin{vmatrix} d_3 & e_3 \\ d_4 & e_4 \end{vmatrix}$$

vanishes, then the pencil contains a conic of the type $by^2 + 2cxz$. Either $c = 0$ and the pencil contains the double line y^2, or $c \neq 0$ and we can take $f_3 = b_3 y^2 + 2xz$, $c_4 = 0$, and also $e_4 = 1$ and $b_3 \neq 0$, since the pencil has no fixed point. The conic

$$\lambda_1 x^2 + 2\lambda_2 xy + \lambda_3 f_3 + f_4$$

is a double line if and only if

$$\lambda_3 b_3 + b_4 - d_4^2 = 0 \qquad \lambda_1 = \lambda_3^2 \qquad \lambda_2 = d_4 \lambda_3 \ .$$

This system admits a solution in $\lambda_1, \lambda_2, \lambda_3$.

We now assume $E \neq 0$. We can take

$$f_3 = b_3 y^2 + 2 c_3 xz + 2 yz \ ,$$
$$f_4 = b_4 y^2 + 2 c_4 xz + z^2 \ .$$

The conic

$$\lambda_1 x^2 + 2\lambda_2 xy + \lambda_3 f_3 + f_4$$

is a double line if and only if

$$\lambda_3^2 - \lambda_3 b_3 - b_4 = 0 \qquad \lambda_1 = (\lambda_3 c_3 + c_4)^2 \qquad \lambda_2 = \lambda_3 (\lambda_3 c_3 + c_4) \ .$$

This system has a solution in $\lambda_1, \lambda_2, \lambda_3$.

1.3 Steiner Surfaces

If $D \neq 0$, we can take

$$f_2 = 2a_2 xy + b_2 y^2 + 2xz$$
$$f_3 = 2a_3 xy + b_3 y^2 + 2yz$$
$$f_4 = 2a_4 xy + b_4 y^2 + z^2 \ .$$

The conic

$$\lambda_1 x^2 + \lambda_2 f_2 + \lambda_3 f_3 + f_4$$

is a double line if and only if

$$\lambda_1 = \lambda_2^2 \quad \lambda_3^2 - \lambda_3 b_3 - \lambda_2 b_2 - b_4 = 0 \quad \lambda_2 \lambda_3 - \lambda_2 a_2 - \lambda_3 a_3 - a_4 = 0 \ .$$

If $b_2 \neq 0$, the system admits a solution which gives a second double line in the pencil. In the case where $b_2 = 0$, replacing z by $a_2 y + z$ allows to write $f_2 = 2xz$ without changing the form of f_3, f_4. The pencil admits no fixed point; thus, b_3 and b_4 do not vanish simultaneously, and the above system has again a solution. q.e.d.

Proposition 6.

There are exactly four types of 3-pencils of conics without fixed point, up to homography:

$$[x^2, y^2, yz, xy + z^2], \quad [x^2, y^2, z^2, x(y+z)]$$
$$[x^2, y^2, z^2, xy], \quad [yz, zx, xy, x^2 + y^2 + z^2] \ .$$

Proof.

By the previous lemma we can set $f_1 = x^2$, $f_2 = y^2$, and

$$f_3 = 2a_3 xy + 2b_3 xz + 2c_3 yz + d_3 z^2$$
$$f_4 = 2a_4 xy + 2b_4 xz + 2c_4 yz + d_4 z^2 \ .$$

The pencil $[f_1, f_2, f_3, f_4]$ admits no fixed point, thus the determinant $\begin{vmatrix} c_3 & d_3 \\ c_4 & d_4 \end{vmatrix}$ is non-zero, and by linear combination we may assume

$$f_3 = 2a_3 xy + 2b_3 xz + 2yz$$
$$f_4 = 2a_4 xy + 2b_4 xz + z^2 \ .$$

$\underline{a_4 \neq 0, \ b_3 = 0, \ b_4 = a_3}$:

We change $a_3 x + z$ in z, $2y$ in y, $a_4 x$ in x, and the pencil is given by $[x^2, y^2, yz, xy + z^2]$.

$\underline{a_4 = 0}$:

The pencil contains the double line $(z + b_4 x)^2$.

$\underline{a_4 \neq 0, \ b_3 \neq 0 \text{ or } b_4 \neq a_3}$:

The conic $\lambda_1 f_1 + \lambda_2 f_2 + f_3 + \lambda_4 f_4$ is a double line if and only if

$$a_4 \lambda_4^2 + \lambda_4 (a_3 - b_4) - b_3 = 0 \quad \lambda_2 \lambda_4 = 1 \quad \lambda_1 = \lambda_4 (a_3 + a_4 \lambda_4)^2$$

This system admits a solution in $\lambda_1, \lambda_2, \lambda_4$.

In the two last cases there exist a third double line independent of the two first ones; We can thus suppose

$$f_3 = z^2 \quad f_4 = axy + bxz + cyz$$

It is easy to check that there are exactly three cases up to homography according to the number of vanishing terms in a, b, c:

$$[x^2, y^2, z^2, xy], \quad [x^2, y^2, z^2, x(y+z)]$$

and $[x^2, y^2, z^2, xy + yz + zx]$ which is equivalent up to homography to

$$[yz, zx, xy, x^2 + y^2 + z^2]$$

q.e.d.

We are now ready to give the classification of Steiner surfaces by elimination of x, y, z:

S1. $(XY - T^2)^2 = 0$

This is a double cone of the second order corresponding to the parametrization $[x^2, y^2, z^2, xy]$.

S2. $(TY - Z^2)^2 - XY^3 = 0$

The singular set of this quartic surface parametrized by $[x^2, y^2, yz, xy + z^2]$ is the line $Y = Z = 0$, image of the line $y = 0$.

S3. $T^4 - 2T^2 X(Y + Z) + X^2(Y - Z)^2 = 0$

This quartic surface is parametrized by $[x^2, y^2, z^2, x(y+z)]$ and its singular set is given by the two coplanar lines $T = X = 0$ and $T = Y - Z = 0$, images of the lines $x = 0$ and $y + z = 0$.

S4. $X^2 Y^2 + Y^2 Z^2 + Z^2 X^2 - XYZT = 0$

The relevant parametrization is $[yz, zx, xy, x^2 + y^2 + z^2]$, and the singular set is constituted by the three non-coplanar lines $X = Y = 0$, $Y = Z = 0$, $Z = X = 0$, meeting in $(0, 0, 0, 1)$ and coming from the three lines $x = 0, y = 0, z = 0$.

The Steiner surfaces are the images of the complex projective plane with a generic projection to 3-dimensional complex projective space composed by the *Veronese embedding* $(x^2, y^2, z^2, yz, zx, xy)$. By restriction in the source and the target to the real parts we can define images of $\mathbb{P}^2$ in the 3-dimensional real projective space $\mathbb{P}^3$.

Consider in particular the generic case S4. We want to deal with the representations of $\mathbb{P}^2$ in $\mathbb{R}^3$, so we have to consider only those images which do not meet a given projective plane in $\mathbb{P}^3$. Up to homography there are two real restrictions of S4 depending on whether the number of real double lines is equal to one or three. In each case we can choose projective coordinates such that the image of $\mathbb{P}^2$ stays in $\mathbb{R}^3$, that means it doesn't meet the plane $T = 0$:

S41. $X^2 Y^2 + Y^2 Z^2 + Z^2 X^2 - XYZT = 0$

This real Steiner surface is called the *Roman surface* parametrized by $[yz, zx, xy, x^2 + y^2 + z^2]$.

S42. $4X^2(X^2 + Y^2 + Z^2 + ZT) + Y^2(Y^2 + Z^2 - T^2) = 0$

This is the *cross-cap* parametrized by $[yz, 2xy, x^2 - y^2, x^2 + y^2 + z^2]$.

1.3 Steiner Surfaces

In the two cases the image of $\mathbb{P}^2$ is a compact subset of $\mathbb{R}^3$ embedded in $\mathbb{P}^3$ by $(X, Y, Z) \mapsto (X, Y, Z, 1)$. At the international congress of mathematicians in Paris in 1900, this was the state of representations of $\mathbb{P}^2$ in $\mathbb{R}^3$ (**Plates 18–22**).

The parametrizations of these two surfaces define two *topological immersions* of $\mathbb{P}^2$ in $\mathbb{R}^3$, that is two locally injective continuous maps. The images of these topological immersions are *ambient isotopic* to the images in $\mathbb{R}^3$ of the Rheinhardt heptahedron and the combinatorial cross-cap (representation 6, 7 sect. 1.2) (Figs. 7, 8). By an *ambient isotopy* between M_0 and M_1 in $\mathbb{R}^3$ we mean a continuous map $\mathbb{R}^3 \times I \xrightarrow{F} \mathbb{R}^3$ so that for every t, $F(.,t)$ is a homeomorphism, $F(.,0) = 1_{\mathbb{R}^3}$ and $F(M_0, 1) = M_1$.

These images of $\mathbb{P}^2$ in $\mathbb{R}^3$ have two main defaults. An algebraic one is that the image of $\mathbb{P}^2$ is included, but not equal to the zero set of its equation:

The singular set of the Roman surface is the union of three non-coplanar double lines meeting in the triple point $(0, 0, 0)$, but the image of $\mathbb{P}^2$ by the parametrization does not reach the six half-lines (**Plates 21, 22**)

$X = Y = 0 \qquad |Z| > 1/2$
$Y = Z = 0 \qquad |X| > 1/2$
$Z = X = 0 \qquad |Y| > 1/2$

In the case of the cross-cap the (real) singular set is the double line $X = Y = 0$. The two other double lines are complex conjugates

$Y - i\sqrt{2}X = 0 \qquad Z + T = 0$
$Y + i\sqrt{2}X = 0 \qquad Z + T = 0$

The triple point of the complex surface, namely $(0, 0, -1)$, is simple for the real parametrization. As in the previous case, the image of $\mathbb{P}^2$ does not contain the two half-lines (**Plate 20**)

$X = Y = 0 \qquad Z > 1$
$X = Y = 0 \qquad Z < 0$

The second default of these images is rather geometric. They do not admit any tangent plane at certain points. These points are the extremities of the previous half-lines.

As a matter of fact, these two defaults come about because of the existence of singularities for the parametrizations, i.e. points at which the jacobian matrix is not of maximal rank. In the next chapter we shall describe these singularities and explain how we can eliminate them in order to smooth the parametrization of $\mathbb{P}^2$.

Chapter 2

The Boy Surface

2.1 Embedded Surfaces and Immersed Surfaces

In order to improve the representations of the real projective plane $\mathbb{P}^2$ in $\mathbb{R}^3$, we have to introduce the notion of a smooth structure on an n-dimensional manifold (sect. 1.1).

An *atlas* A on the n-manifold M will be a set of homeomorphisms h_i from open subsets U_i of M onto open subsets of the half-space $(\mathbb{R}^n)_+ = \{(x_1, \ldots, x_n) \in \mathbb{R}^n : x_n \geq 0\}$, such that $M = \bigcup_{i \in I} U_i$, and such that for every i and j the homeomorphism

$h_i (U_i \cap U_j) \xrightarrow{h_j h_i^{-1}} h_j (U_i \cap U_j)$ is the restriction of a C^∞-map from $\mathbb{R}^n$ to itself.

The *smooth structure on M induced by A* is by definition the atlas containing A which is maximal with respect to inclusion; with this structure M becomes a *smooth manifold*. If $U \to hU$ is a homeomorphism of the smooth structure and the point x belongs to U, then we say (U, h) a *local coordinate system at x*.

Example 1.

On the manifolds $\mathbb{R}^n$ and $(\mathbb{R}^n)_+$ we can define a smooth structure by a trivial atlas, that is an atlas consisting of a single local coordinate system. For an n-manifold without boundary it is possible to define the smooth structure in such a way that the images of the local coordinate systems are open subsets of $\mathbb{R}^n$ instead of $(\mathbb{R}^n)_+$.

Example 2.

An open subset U of a smooth manifold M has a canonical smooth structure given by taking the restriction of the smooth structure on M.

Example 3.

Let $\bar{B}^n$ the closed unit ball of $\mathbb{R}^n$. We denote by e a number equal to 1 or -1, and by U_e the open subset of $\bar{B}^n$ equal to the complement of $(0, \ldots, 0, -e)$. We define the homeomorphism

$$h_e(y_1, \ldots, y_n) = (1 + ey_n)^{-1} \cdot (y_1, \ldots, y_{n-1}, (1 - y_1^2 - \ldots - y_n^2)^{1/2})$$

from U_e onto $(\mathbb{R}^n)_+$. We have

$$h_{-e} h_e^{-1}(x_1, \ldots, x_n) = (x_1^2 + \ldots + x_n^2)^{-1} \cdot (x_1, \ldots, x_n),$$

so $h_{-e} h_e^{-1}$ is a C^∞-homeomorphism from $(\mathbb{R}^n)_+ \setminus \{0\}$ to itself.

2.1 Embedded Surfaces and Immersed Surfaces

Thus (U_1, h_1) and (U_{-1}, h_{-1}) constitute an atlas on $\bar{B}^n$. We have defined the canonical smooth structure on $\bar{B}^n$.

Let x be a point of the boundary ∂M of the n-manifold M, and (U, h) be a coordinate system at x, then we have

$$h(\partial M \cap U) \subseteq \mathbb{R}^{n-1} .$$

The family of coordinate systems $(\partial M \cap U, h)$ on ∂M defines a smooth structure. The boundary of a smooth manifold thus has a canonically smooth manifold structure.

For instance, the canonical smooth structure on the sphere S^{n-1} is the one inherited from that of $\bar{B}^n$ defined in example 3.

Let $X \xrightarrow{p} B$ be a covering (sect. 1.2) where X is a smooth manifold. It is possible to construct an atlas on X such that for each local coordinate system (U, h) the composite ph^{-1} is a homeomorphism. Then the inverse maps of these homeomorphisms define an atlas and a smooth structure on the base B. This smooth structure does not depend on the chosen atlas on X.

Example 4.

If we define the *n-dimensional real projective space* $\mathbb{P}^n$ as the quotient of the sphere S^n by the antipodal relation, then the quotient surjection is the projection of a 2-sheeted covering of $\mathbb{P}^n$ by S^n. By the previous construction we can provide $\mathbb{P}^n$ with a smooth structure.

Example 5.

Prop. 7 sect. 1.2 proves that every closed surface can be covered by the sphere S^2 or the plane $\mathbb{R}^2$. Thus, each closed surface can be provided with a smooth structure.

A map $M \xrightarrow{f} N$ between two smooth manifolds will be said to be *differentiable*, resp. *r-times differentiable, of class C^r*, or *smooth* (that is of class C^∞) if for every pair of local coordinate systems (U, h) in M and (V, k) in N verifying $f(U) \subseteq V$, the map kfh^{-1} has the same property. Each of these properties is preserved by composition. A C^r-*diffeomorphism* is a homeomorphism of class C^r for which the inverse is also of class C^r. For instance, a coordinate system on a smooth manifold defines a smooth diffeomorphism. Note that a map from $\bar{B}^n$ or $(\mathbb{R}^n)_+$ is smooth if and only if it can be extended to a C^∞-map defined on $\mathbb{R}^n$.

Note that the following proposition (for a proof see for instance [HI]) does not generalize to higher dimensions:

Proposition 1.

On a curve or a surface there is exactly one smooth structure up to smooth diffeomorphism.

Example 6.

Consider the map defined by

$$f(x, y) = 2^{-1/2} \cdot (x^2 + y^2)^{1/4} \cdot (((x^2 + y^2)^{1/2} + x)^{1/2}, ((x^2 + y^2)^{1/2} - x)^{1/2}) .$$

It is a homeomorphism from $(\mathbb{R}^2)_+$ onto $\mathbb{R}_+ \times \mathbb{R}_+$. The inverse map gives an atlas and hence a smooth structure on $\mathbb{R}_+ \times \mathbb{R}_+$. The projection of $\mathbb{R}_+ \times \mathbb{R}_+$ onto the first factor is clearly not differentiable. We can prove that there exists only one smooth structure on $\mathbb{R}_+ \times \mathbb{R}_+$ up to smooth diffeomorphism. Thus we see that for manifolds with boundary it is not possible to define a product smooth structure, which makes the projections smooth.

Proposition 2.

If M and N are two smooth manifolds, the boundary of N being empty, then there exists an unique smooth structure on the product manifold verifying the universal property of the product, namely: the two projections $M \times N \xrightarrow{p} M$ and $M \times N \xrightarrow{q} N$ are smooth, and for every pair of smooth maps $L \xrightarrow{p'} M$ and $L \xrightarrow{q'} N$ there is exactly one smooth map $L \xrightarrow{r} M \times N$ such that $pr = p'$ and $qr = q'$.

The boundary of $M \times N$, that is $\partial M \times N$, is a smooth manifold without boundary.

Proof.

The key to the proof lies in the fact the product $\mathbb{R}_+ \times \mathbb{R}$ is a smooth manifold in a trivial way. q.e.d.

Example 7.

The annulus $S^1 \times I$ and the product $S^1 \times S^1$ can be provided with the product smooth structures. In the second case we obtain the smooth torus $\mathbb{T}^2$.

As we saw in sect. 1.1, it is possible to represent the sphere S^2 and the torus $\mathbb{T}^2$ in $\mathbb{R}^3$ by a topological embedding, and using the connected sum, such a topological embedding in $\mathbb{R}^3$ exists for every connected orientable surface. If we want to make these representations more regular, for instance by imposing the existence of a tangent plane, we need to introduce the notion of C^r-embedding (**Plates 7, 8, 10, 12**).

First of all we have to define the *rank* of a differentiable map $M \xrightarrow{f} N$ between two smooth manifolds. Let (U, h) be a coordinate system at x on M and (V, k) be a coordinate system at $f(x)$ on N so that $f(U) \subseteq V$. Then the rank of the jacobian matrix of kfh^{-1} at x is independent of the chosen coordinate systems. It is called the rank of f at x and it is denoted by $rkf(x)$. By definition the map f shall be a C^r-*embedding* when it is a topological embedding of class C^r whose rank is constant and equal to $\dim M$ ($r \geq 1$).

A way to recognize the image of a smooth embedding is the following:

Proposition 3.

Let M and N be two smooth manifolds without boundary, and $M \xrightarrow{f} N$ be a smooth map whose rank is equal to the constant k in a neighborhood of the inverse image $f^{-1}(y)$ of a point y in N. Then $f^{-1}(y)$ is the image of a smooth manifold without boundary of dimension $\dim M - k$ by a smooth embedding.

The proof, for which we do not give details, is a direct application of the implicit function theorem (see e.g. [HI]).

2.1 Embedded Surfaces and Immersed Surfaces

Example 8.

The map $f(x, y, z) = x^2 + y^2 + z^2 - 1$ from $\mathbb{R}^3$ to $\mathbb{R}$ is of rank 1 in a neighborhood of $f^{-1}(0)$. Hence, $f^{-1}(0)$ is the image of a smooth surface, namely the smooth sphere S^2, by a smooth embedding **(Plate 1)**.

A useful tool for computing the Euler characteristic of a smooth surface is the Morse theorem. Before stating it we have to introduce the notion of a *critical point* of a C^2-map $M \xrightarrow{f} \mathbb{R}$. This is a point on the smooth manifold M at which the rank of f vanishes. Let x be a critical point of f and (U, h) be a coordinate system at x on M. Then the *Hessian quadratic form* of fh^{-1}, that is the quadratic term in the Taylor expansion of fh^{-1}, is defined (up to equivalence) independently of the coordinate system. The critical point will be called *non-degenerate* if the Hessian quadratic form is nondegenerate, and we define the *index* of the critical point to be the number of negative eigenvalues of the matrix of the Hessian quadratic form.

We shall say f is a *Morse function* on a smooth manifold M without boundary, if it is a C^2-map into $\mathbb{R}$ for which all the critical points are non-degenerate. Here is a form of the Morse theorem for a surface M without boundary (this form is probably due to W. Boy [BOY]):

Proposition 4.

If f is a Morse function on the smooth surface M without boundary, then the number of critical points is finite, and

$$\chi(M) = n_0 - n_1 + n_2$$

where n_i is the number of critical points of index i.

Moreover, the number of components of M is less than or equal to min $\{n_0, n_2\}$.

According as the index of the critical point is 0, 1 or 2, it is called a *minimum*, a *saddle*, or a *maximum* for f. If $M \xrightarrow{f} \mathbb{R}^3$ is a smooth embedding for the smooth surface M in $\mathbb{R}^3$, we shall refer to fM as an *embedded surface;* we note that there exists exactly one smooth structure on fM which makes f a smooth diffeomorphism.

Example 9.

Consider the map $\mathbb{R}^3 \xrightarrow{f} \mathbb{R}$ defined by

$$f(x, y, z) = g^2(x, y) + z^2 - 1/2$$

where

$$g(x, y) = y^2 + (x^2/p^2 - 1) \left(\sum_{k=0}^{[\frac{p-1}{2}]} (-1)^k \binom{p}{2k+1} (x/p)^{p-2k-1}(1 - x^2/p^2)^k \right)^2.$$

Here p is an integer greater or equal to 1, and $[x]$ denotes the greatest integer less than or equal to x.

We shall need the following two relations

(i) $\quad g(p\cos u, y) = y^2 - \sin^2 pu$

(ii) $\quad g(\pm p\cosh u, y) = y^2 + \sinh^2 pu$.

We denote by M the subset $f^{-1}(0)$ in $\mathbb{R}^3$. Assuming that the point (x, y, z) on M verifies

$$\partial_y f = \partial_z f = f = 0,$$

we obtain

$$y = z = 0 \quad \text{and} \quad |g(x, 0)| = 1/\sqrt{2}.$$

By relations (i) and (ii) we see that $|x|$ is not equal to p:

$|x| < p$

We write $x = p\cos u$ where u belongs to $]0, \pi[$; then using (i) we have

$$g(x, 0) = -\sin^2 pu = -1/\sqrt{2}.$$

Differentiating with respect to u we have

$$-p\sin u \cdot \partial_x g(x, 0) = -2p\sin pu \cdot \cos pu\ ;$$

since $\sin u \neq 0$, we have

$$\partial_x g(x, 0) = 2\sin pu \cdot \cos pu \cdot (\sin u)^{-1}.$$

If $\cos pu = 0$, then $|g(x, 0)| = 1 \neq 1/\sqrt{2}$; hence $\cos pu \neq 0$. If $\sin pu = 0$, then $|g(x, 0)| = 0 \neq 1/\sqrt{2}$; hence $\sin pu \neq 0$. Finally,

$$\partial_x g(x, 0) \neq 0 .$$

$|x| > p$

We write $x = \pm p\cosh u$ with $u > 0$, and from (ii) we deduce that

$$g(x, 0) = \sinh^2 pu .$$

Differentiating with respect to u we have

$$\partial_x g(x, 0) = \pm 2\sinh pu \cdot \cosh pu \cdot (\sinh u)^{-1} \neq 0 .$$

We conclude that the rank of f is nonzero on M, hence it is equal to 1 in a neighborhood of M in $\mathbb{R}^3$. By propostion 3, M is the image of a smooth embedding of a smooth surface without boundary. We are now going to introduce a Morse function on this embedded surface so as to compute its Euler characteristic.

On M we consider the function $h(x, y, z) = x$. The critical points of h on M verify

$$y = z = 0 \quad \text{and} \quad |g(x, 0)| = 1/\sqrt{2} .$$

As previously, if (x, y, z) is a critical point of h on M, we have $|x| \neq p$:

2.1 Embedded Surfaces and Immersed Surfaces

$|x| < p$

We write $x = p \cos u$ with $u \in {]}0, \pi[$. Then, using (i), the equation of M is equivalent in a neighborhood of $(x, 0, 0)$, to

$$u = (1/p) \operatorname{Arcsin} ((1/2 - z^2)^{1/2} + y^2)^{1/2} + k\pi/p$$

This formula gives a coordinate system on M at $(x_k, 0, 0)$ by

$$(y, z) \mapsto (p \cos u, y, z)$$

where $x_k = p \cos u_k$ and $u_k = (\operatorname{Arcsin}(2^{-1/4}) + k\pi)/p$. The Hessian quadratic form at $(x_k, 0, 0)$ is given by

$$2^{-3/4} \cdot (1 - 1/\sqrt{2})^{-1/2} \sin u_k \cdot (-\sqrt{2} y^2 + z^2).$$

Thus, h has $2p$ saddles at the points $(x_k, 0, 0)$ with $0 \leq k \leq 2p - 1$.

$|x| > p$

We write $x = \pm p \cosh u$ with $u > 0$. Then, using (ii), the equation of M is equivalent in a neighborhood of $(x, 0, 0)$, to

$$u = (1/p) \operatorname{Arcsinh} ((1/2 - z^2)^{1/2} - y^2)^{1/2}$$

This formula gives a coordinate system on M at $(x_1, 0, 0)$ or $(x_{-1}, 0, 0)$ by

$$(y, z) \mapsto (\pm p \cosh u, y, z)$$

where $x_1 = p \cosh v$, $x_{-1} = -p \cosh v$ and $v = (1/p) \operatorname{Arcsinh}(2^{-1/4})$. The Hessian quadratic form at $(x_1, 0, 0)$ or $(x_{-1}, 0, 0)$ is given by

$$-(\pm 2^{-3/4})(1 + 1/\sqrt{2})^{-1/2} \sinh v \cdot (\sqrt{2} y^2 + z^2).$$

Thus h has one maximum and one minimum situated respectively at $(x_1, 0, 0)$ and $(x_{-1}, 0, 0)$.

We conclude by proposition 4 that M is a connected surface of Euler characteristic

$$\chi(M) = 1 - 2p + 1.$$

Thus M is a closed embedded surface of genus p. Prop. 7 sec. 1.1 proves that this surface is orientable (**Plates 10–12**).

Example 10.

Consider a smooth embedding $M \xrightarrow{f} \mathbb{R}^3$ of a closed orientable surface M, and let (U, h) be a coordinate system at x on M. In $hU \subseteq (\mathbb{R}^2)_+$ we consider n open disjoint disks $D_1, \ldots, D_n$ whose boundaries are n disjoint circles not meeting the boundary of $(\mathbb{R}^2)_+$. Clearly $N = M \setminus h^{-1}(D_1 \cup \ldots \cup D_n)$ is a smooth surface with boundary, and f restricts to a smooth embedding of N in $\mathbb{R}^3$. By this procedure we see that any orientable surface can be smoothly embedded in $\mathbb{R}^3$.

Example 11.

In example 7 sec. 1.2 we defined an 2-sheeted covering of the Möbius strip by the annulus. By this covering the smooth structure of the annulus (example 7) induces the smooth structure of the Möbius strip M.

Following an idea of L. Siebenmann we can construct an explicit smooth embedding from M into $\mathbb{R}^3$ such that the boundary is a circle in $\mathbb{R}^3$.

Consider the map $[0, \pi] \times [0, \pi] \xrightarrow{f} S^3 \subseteq \mathbb{R}^4$ defined by

$$f(t, u) = (\cos 2t \cdot \sin u, \sin 2t \cdot \sin u, \cos t \cdot \cos u, \sin t \cdot \cos u).$$

This map factorizes through the quotient map $[0, \pi] \times [0, \pi] \xrightarrow{p} M$ where M is the Möbius strip and $p(0, u) = p(\pi, \pi - u)$, giving a smooth embedding $M \xrightarrow{g} S^3$. The image of the boundary ∂M is the great circle of S^3 with equation $X = Y = 0$ and $Z^2 + T^2 = 1$. After the coordinate change

$$X' = X \quad Y' = (Y - T)/\sqrt{2} \quad Z' = Z \quad T' = (Y + T)/\sqrt{2}$$

we project stereographically from the pole $(0, 0, 0, 1)$ onto $\mathbb{R}^3$. We obtain a smooth embedding $M \to \mathbb{R}^3$ such that the image of the boundary of M is the circle $x = 0$ and $(y + 1)^2 + z^2 = 2$ in $\mathbb{R}^3$. **(Plate 13)**.

To investigate the smooth representation of non-orientable closed surfaces in $\mathbb{R}^3$, we introduce the notion of C^r-immersion (smooth means C^∞).

Let $M \xrightarrow{f} N$ be a C^r-map between two smooth manifolds. We shall say f is a C^r-immersion if its rank is constant and equal to $\dim M$. In particular, f must be a topological immersion (sec. 1.3). Contrary to an embedding, there can be *self-intersection points*, that is points having at least two preimages. The set of these points is called the *self-intersection set*.

Example 12.

We are going to exhibit a non-trivial smooth immersion of the smooth torus $\mathbb{T}^2$ in $\mathbb{R}^3$. The image of this immersion is traced out by a figure of eight whose double point describes a circle. The plane of the figure of eight is always orthogonal to the circle, and as the double point makes a complete revolution, the figure of eight makes a turn in its plane.

We begin to choose the parametrization of the figure of eight in the moving plane using the frame (A, Y, Z) where A runs along the circle $(x, y, z) = (4 \cos u, 4 \sin u, 0)$. The formulae for this coordinate change are

$$x = 4 \cos u + \cos u \cdot Y$$
$$y = 4 \sin u + \sin u \cdot Y$$
$$z = Z$$

We now give a parametrization of a figure of eight *(the lemniscate of Bernoulli)*

$$r = \sin 2t \cdot (\cos^4 t + \sin^4 t)^{-1}$$
$$Y = r$$
$$Z = r \cos 2t$$

2.1 Embedded Surfaces and Immersed Surfaces

In order to make the figure of eight rotate about A, we put

$$Y = r(\cos u - \sin u \cdot \cos 2t)$$
$$Z = r(\sin u + \cos u \cdot \cos 2t).$$

Finally, we define the smooth map $[0, \pi] \times [0, 2\pi] \xrightarrow{f} \mathbb{R}^3$ by

$$f(t, u) = \begin{bmatrix} 4\cos u + r\cos 2u \cdot (\cos u - \sin u \cdot \cos 2t) \\ 4\sin u + r\sin 2u \cdot (\cos u - \sin u \cdot \cos 2t) \\ r \cdot (\sin u + \cos u \cdot \cos 2t) \end{bmatrix}$$

To compute the rank of f we use the following relations

$$\sin u \cdot \partial_u x - \cos u \cdot \partial_u y = -r(\cos u - \sin u \cdot \cos 2t) - 4 < 0$$
$$\sin u \cdot \partial_t x - \cos u \cdot \partial_t y = 0$$
$$\cos u \cdot \partial_t x + \sin u \cdot \partial_t y = r' \cdot \cos u + \sin u \cdot (2r \sin 2t - r' \cos 2t)$$
$$\partial_t z = r' \cdot \sin u - \cos u \cdot (2r \sin 2t - r' \cos 2t)$$

The left hand sides of the two last equations vanish simultaneously if and only if $r' = r = 0$, what is impossible. Hence, there is a linear combination of 2×2-minors of the jacobian matrix of f which does not vanish. It proves the rank of f is equal to 2, and so f is a smooth immersion.

Bearing in mind the fact that $f(0, u) = f(\pi, u)$ and $f(t, 0) = f(t, 2\pi)$ we see that f factorizes through the quotient map $[0, \pi] \times [0, 2\pi] \to \mathbb{T}^2$ to give the wanted smooth immersion from $\mathbb{T}^2$ into $\mathbb{R}^3$.

Note that the self-intersection set is given by the circle described by the double point of the figure of eight.

Example 13.

Provided with the atlas given by the two coordinate systems

$$\mathbb{C} \to \mathbb{R}^2 \qquad\qquad \mathbb{C}^* \cup \{\infty\} \to \mathbb{R}^2$$
$$z \mapsto (\operatorname{Re} z, \operatorname{Im} z) \qquad\qquad z \mapsto |z|^{-2} \cdot (\operatorname{Re} z, -\operatorname{Im} z)$$

the complex projective line $\mathbb{C} \cup \{\infty\}$ becomes a smooth surface diffeomorphic to the sphere S^2 by stereographic projection (rep.8 sec.1.2). A rational function f from $\mathbb{C} \cup \{\infty\}$ to itself is smooth. We write $f(z) = p(x, y) + iq(x, y)$ where $p = \operatorname{Re} f$, $q = \operatorname{Im} f$, $x = \operatorname{Re} z$ and $y = \operatorname{Im} z$. The Cauchy relations

$$\partial_y p = -\partial_x q \qquad\qquad \partial_y q = \partial_x p$$

prove if z is not a pole of f then $|f'(z)|^2$ is equal to the Jacobian determinant of $(p(x, y), q(x, y))$, whence we conclude the rank of f is equal to 0 or 2. In fact this property of the rank holds for every z in $\mathbb{C} \cup \{\infty\}$ since the group of homographies acts as a transitive set of smooth diffeomorphisms of $\mathbb{C} \cup \{\infty\}$ to itself.

Example 14.

The first immersion of a non-orientable closed surface in $\mathbb{R}^3$ was discovered by F. Klein in 1882, and the shape of its image justifies the name of Klein bottle $\mathbb{K}$ (sec. 1.1) (Plates 15, 16).

In fact Klein constructed by geometrical methods a subset of $\mathbb{R}^3$ which is the image of the smooth closed non-orientable surface of genus 2 by a smooth immersion, but he did not give any explicit parametrization. If the lemniscate of Bernoulli of example 12 makes half a turn instead of a turn as the double point makes a complete revolution, we obtain a smooth immersion of $\mathbb{K}$ in $\mathbb{R}^3$ **(Plate 9)** [PI].

B. Morin extended Siebenmann's idea for the Möbius strip (example 11) to construct a Klein bottle generated by circles for which the self-intersection set is also a circle. The parametrization is given by the same formulae as example 11, but we extend f to a map $[0, \pi] \times [-\pi/2, 3\pi/2] \xrightarrow{f} S^3$. The quotient of the square $[0, \pi] \times [-\pi/2, 3\pi/2]$ by the equivalence relation

$$(0, u) \sim (\pi, \pi - u) \quad \text{and} \quad (t, -\pi/2) \sim (t, 3\pi/2)$$

gives a 2-complex of characteristic $1 - 2 + 1 = 0$. This complex contains a Möbius strip, namely the restriction of the quotient to $[0, \pi] \times [0, \pi]$. In fact, it is the Klein bottle $\mathbb{K}$ and f induces a smooth map $\mathbb{K} \xrightarrow{g} S^3$. The rank of f is maximal, and hence so is that of g; thus g is a smooth immersion. After the stereographic projection indicated in example 11, we obtain the following immersion of $\mathbb{K}$ in $\mathbb{R}^3$ denoted by $(x, y, z) = h(t, u)$ where $(t, u) \in [0, \pi] \times [-\pi/2, 3\pi/2]$:

$$w = (\sin t \cdot \cos u + \sin 2t \cdot \sin u)/\sqrt{2}$$
$$x = \cos 2t \cdot \sin u/(1 - w)$$
$$y = (\sin 2t \cdot \sin u - \sin t \cdot \cos u)/(\sqrt{2}(1 - w))$$
$$z = \cos t \cdot \cos u/(1 - w).$$

The image of $\mathbb{K}$ in S^3 is the zero set of the polynomial

$$Y(T^2 - Z^2) + 2XZT.$$

Thus, using the formulae for coordinate change of example 11, we see that the image of $\mathbb{K}$ in $\mathbb{R}^3$ is exactly the zero set of the polynomial of degree six

$$(x^2 + y^2 + z^2 + 2y - 1)((x^2 + y^2 + z^2 - 2y - 1)^2 - 8z^2) + 16xz(x^2 + y^2 + z^2 - 2y - 1).$$

The singular set of this algebraic surface is the circle

$$z = 0 \quad \text{and} \quad x^2 + (y - 1)^2 = 2.$$

In 1900 D. Hilbert, thinking it was impossible to immerse the projective plane in $\mathbb{R}^3$, asked his student W. Boy to prove this conjecture. In his thesis in 1901, Boy constructed a smooth immersion of $\mathbb{P}^2$ in $\mathbb{R}^3$ [BOY]. As in the case of the Klein construction, this immersion was not explicit but arose from methods of analysis situs. The combinatorial Boy surface of representation 9 sec. 1.2 is ambient isotopic to the model constructed by W. Boy. Let us use the term *Boy immersion* to describe any smooth immersion of $\mathbb{P}^2$ in $\mathbb{R}^3$.

The importance of the Boy immersion is illustrated by a theorem of U. Pinkall. We need to define the notion of *regular homotopy* between two smooth immersions $M \xrightarrow{f_0} \mathbb{R}^3$ and $M \xrightarrow{f_1} \mathbb{R}^3$ of a smooth closed surface M into $\mathbb{R}^3$; this is a smooth map $M \times I \xrightarrow{F} \mathbb{R}^3$

2.2 Parametrization of the Boy Surface by 3 Polynomials

such that $F(.\,,0) = f_0$, $F(.\,,1) = f_1$ and $F(.\,,t)$ is a smooth immersion for every t in I. The immersions f_0 and f_1 are said to be equivalent whenever there exists a smooth diffeomorphism $M \xrightarrow{g} M$ such that f_0 and $f_1 \circ g$ are regularly homotopic. The equivalence classes are called *immersed closed surfaces*. Note that the *connected sum* of two smooth closed surfaces S and S' can be made smoothly in such a way as to obtain a new smooth closed surface denoted by $S \# S'$ which is homeomorphic to the one of prop. 8 sec. 1.1 (see [HI]).

Theorem 5 [PI].

The operation # can be extended to immersed closed surfaces so that it yields a countable commutative semigroup generated by the standard embedding T of the torus (example 5 sec. 1.1), the non-trivial smooth immersion $\bar{T}$ of the torus of example 12, the Boy immersion B and its mirror image $\bar{B}$ (in $\mathbb{R}^3$) (Fig. 19), with the following relations:

$$T\#T = \bar{T}\#\bar{T} \qquad B\#B\#B\#B = \bar{B}\#\bar{B}\#\bar{B}\#\bar{B} \qquad T\#B = B\#B\#\bar{B}$$
$$T\#B = \bar{B}\#\bar{B}\#\bar{B} \qquad T\#\bar{B} = B\#B\#B \qquad T\#\bar{B} = B\#\bar{B}\#\bar{B}$$

This theorem shows the importance of the Boy immersion and the next section shall be devoted to some attempts to parametrize its image.

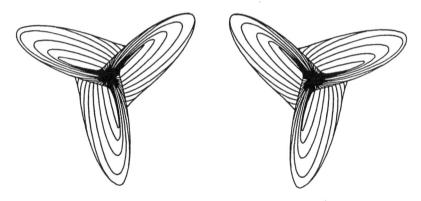

Fig. 19 Direct and inverse Boy surfaces.

2.2 Parametrization of the Boy Surface by Three Polynomials

Since the discovery of Boy several attempts have been made to give explicit parametrizations of the real projective plane in $\mathbb{R}^3$. Before describing some of them, we shall define the geometrical object described by Boy which we will call the *direct Boy surface* to distinguish it from its mirror image which will be called *opposite*.

We start from a C^1-immersion $S^1 \xrightarrow{h} \mathbb{R}^3$ with precisely one triple point at which the three oriented tangents form a righthanded frame in $\mathbb{R}^3$. The three preimages of the triple point separate S^1 into three arcs whose images under h are three circles topologically embedded in $\mathbb{R}^3$. We require these three circles should be the boundaries of three disks (sec. 1.1) in $\mathbb{R}^3$, any two of which have a unique point of intersection at the triple point.

The image of such an immersion is called a *direct three-bladed propeller*, and is unique up to C^1-ambient isotopy. We define a *C^r-ambient isotopy* between two subsets M_0 and M_1 of a smooth manifold N without boundary to be a C^r-map $N \times I \xrightarrow{F} N$ such that for every t, $F(.,t)$ is a C^r-diffeomorphism from N into itself, $F(.,0) = 1_N$ and $F(M_0, 1) = M_1$. In the case mentioned above $N = \mathbb{R}^3$.

We define a canonical model h of the direct three-bladed propeller as follows: using the standard coordinate system in $\mathbb{R}^3$, we consider the points

$A = (1, 0, 0)$ $B = (0, 1, 0)$ $C = (0, 0, 1)$
$A' = (-1, 0, 0)$ $B' = (0, -1, 0)$ $C' = (0, 0, -1)$

Let α be the directed arc formed by the sum of the line segment OB, the arc of the circle in the plane yOz with center $(0, 1, -1)$ starting point B and endpoint C' (we take the longer of the two arcs), and the line segment $C'O$. Now complete the propeller by adding the directed arcs β and γ in that order which are obtained by rotating α through $2\pi/3$ and $4\pi/3$ about the axis $x = y = z$ directed by the vector $(-1, -1, -1)$. There exists a C^1-immersion of S^1 in $\mathbb{R}^3$ whose image is the canonical model h which is the support of $\alpha \beta \gamma$ (Fig. 20).

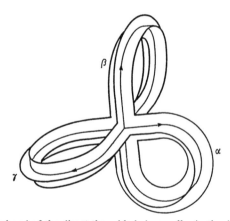

Fig. 20 Neighborhood of the direct three-bladed propeller in the direct Boy surface.

Let H' denote the set of points in the plane yOz at distance less than or equal to $1/3$ from the support of α, and let H'' consist of those points lying on the cylinder generated by α with axis parallel to Ox, at distance less than or equal to $1/3$ from the support of α. We shall denote by H the union of H', H'' and their images under the two rotations defined above **(Plates 46, 47)**.

The boundary of H consists of four connected components of which three bound disks L, M and N lying in coordinate planes; the fourth bounds a disk D in $\mathbb{R}^3$ whose interior is disjoint from $H \cup L \cup M \cup N$.

2.2 Parametrization of the Boy Surface by 3 Polynomials

The representation 9 sec. 1.2 shows a topological realisation of the disk D, and Boy's thesis shows that this construction can be carried out with respect to the class C^1. We define a *direct Boy surface* to be the image of any C^1-immersion of $\mathbb{P}^2$ in $\mathbb{R}^3$ which becomes C^1-ambient isotopic to $H \cup L \cup M \cup N$ when we remove a suitable disk.

The need was felt to explicitly parametrize the Boy surface as soon as it was discovered because of the belief that the existence of a geometrical entity should be confirmed by giving equations. The plans of F. Schilling in 1924 [SC] for the construction of a wire model seem to have been motivated by this worry, but they did not lead to any explicit formulae (Fig. 21).

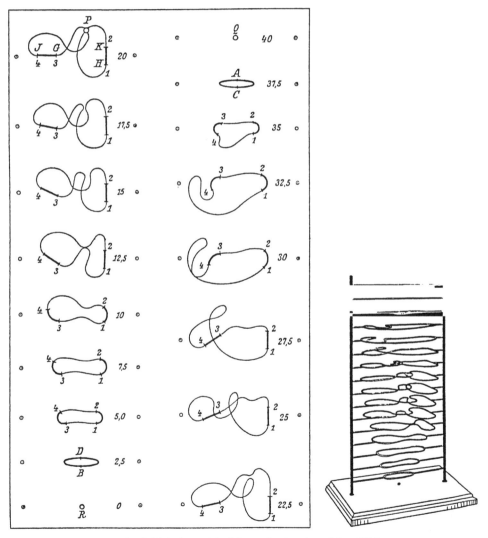

Fig. 21 Schilling's model of the projective plane (after [SC]).

Example 4 sec. 2.1 shows that an immersion of $\mathbb{P}^2$ in $\mathbb{R}^3$ is given by an immersion of the sphere S^2 which is *invariant under the antipodal action,* in other words the images of antipodal points coincide. Such a C^r-immersion can be obtained by a C^r-mapping $\mathbb{R}^3 \setminus (0,0,0) \xrightarrow{f} \mathbb{R}^3$ which is *homogeneous of even degree,* i.e. there exists an integer n such that

$$f(tx, ty, tz) = t^{2n} \cdot f(x, y, z) \quad \text{for all nonzero } t$$

and such that the mapping with target $\mathbb{R}^4$ obtained from f by setting the fourth component equal to the equation of the sphere $x^2 + y^2 + z^2 - 1$, is a C^r-immersion i.e. its rank is three.

As a result of this argument H. Hopf conjectured that such an f exists with polynomial components. We shall show that this construction is possible with polynomials of degree four contrary to the conjecture mentioned in Hopf [HO].

The first immersion of $\mathbb{P}^2$ in $\mathbb{R}^3$ given by algebraic parametrization was found by B. Morin [MO2] in 1978 as the central step in an eversion of the sphere. Here is the corresponding mapping:

$$f(x, y, z) = r^{-1}(x, y, z) \begin{bmatrix} f_1(x, y, z) \\ f_2(x, y, z) \\ s^{-1}(x, y, z)(x^2 + y^2 + z^2)^3 \cdot f_3(x, y, z) \end{bmatrix}$$

where

$$r(x, y, z) = (x^2 + y^2)(3x^4 - 6x^2 y^2 - y^4 + 2zx^3 + 2zxy^2)$$
$$s(x, y, z) = 2(x^2 + y^2 + z^2)^3 + y(y^2 - 3x^2)(z^3 + zx^2 + zy^2 + x^2 - 3xy^2)$$
$$f_1(x, y, z) = 8y(x^2 + y^2)(4xy^2 - x^2 z - y^2 z)$$
$$f_2(x, y, z) = -5x^6 + 57x^4 y^2 - 63x^2 y^4 + 3y^6 - 12x^5 z + 24x^3 y^2 z + 36xy^4 z$$
$$\qquad - 4x^4 z^2 - 8x^2 y^2 z^2 - 4y^4 z^2$$
$$f_3(x, y, z) = 2(13x^6 - 33x^4 y^2 + 87x^2 y^4 + 5y^6 + 12x^5 z - 24x^3 y^2 z - 36xy^4 z$$
$$\qquad + 4x^4 z^2 + 8x^2 y^2 z^2 + 4y^4 z^2)$$

This is indeed a C^1-immersion but fails at one point to be C^2 and it has not been established whether its image is indeed the Boy surface in the sense defined above **(Plate 33)**.

Let us mention an experimental attempt by Petit and Souriau in 1981 to parametrize the Boy surface. This was based on a wire model produced by the sculptor Max Sauze in which the surface is generated by ovals passing through a fixed point, which Petit and Souriau took to be ellipses [PE]. Here is the resulting parametrization:

$$a = 10 + 1.41 \sin(6t - \pi/3) + 1.98 \sin(3t - \pi/6)$$
$$b = 10 + 1.41 \sin(6t - \pi/3) - 1.98 \sin(3t - \pi/6)$$
$$c = a^2 - b^2 \quad d = (a^2 + b^2)^{1/2} \quad e = (\pi/8) \sin 3t$$
$$f = c/d + a \cos u - b \sin u \quad g = d + a \cos u + b \sin u$$

$$X = 3.3 (f \cos t - g \sin e \cdot \sin t)$$
$$Y = 3.3 (f \sin t + g \sin e \cdot \cos t) \qquad t \in [0, \pi] \quad u \in [0, 2\pi]$$
$$Z = 4 (g \cos e - 10)$$

2.2 Parametrization of the Boy Surface by 3 Polynomials

It seems incongruous to seek to calculate the rank of this mapping to discover whether it is indeed an immersion, and although it is pictorially satisfactory, the authors do not give any indication as to how to prove that this is a parametrization of the Boy surface **(Plates 31, 32)**.

In his thesis Boy suggests a polynomial approximation method for obtaining a parametrization of his surface; unfortunately this programme could not be carried out until the era of the computer. This procedure was put into effect by J. F. Hughes (private communication to B. Morin in 1985) who obtained several parametrizations by polynomials of degree eight which, like the preceding formulae, do not seem to be amenable to a computation of the rank. Here is one of these empirical formulae **(Plate 34)**:

$$f(x, y, z) = \begin{bmatrix} g(x, y, z) \\ g(y, z, x) \\ g(z, x, y) \end{bmatrix}$$

where

$$\begin{aligned} g(x, y, z) = &-0.77\, x^8 + 0.36\, x^7 y + 0.17\, x^6 y^2 + 1.09\, x^5 y^3 + 0.62\, x^4 y^4 + 0.34\, x^3 y^5 \\ &+ 0.74\, x^2 y^6 + 0.85\, xy^7 + 0.25\, y^8 - 0.25\, x^7 z + 0.94\, x^6 yz + 0.94\, x^5 y^2 z \\ &+ 2.93\, x^4 y^3 z - 2.97\, x^3 y^4 z + 6.39\, x^2 y^5 z - 0.57\, xy^6 z + 1.18\, y^7 z \\ &- 1.15\, x^6 z^2 + 0.23\, x^5 yz^2 + 3.48\, x^4 y^2 z^2 + 6.30\, x^3 y^3 z^2 + 4.45\, x^2 y^4 z^2 \\ &+ 1.24\, xy^5 z^2 + 1.77\, y^6 z^2 + 0.58\, x^5 z^3 + 3.21\, x^4 yz^3 - 5.16\, x^3 y^2 z^3 \\ &+ 7.32\, x^2 y^3 z^3 + 2.58\, xy^4 z^3 + 1.79\, y^5 z^3 + 2.11\, x^4 z^4 + 2.67\, x^3 yz^4 \\ &+ 8.73\, x^2 y^2 z^4 - 0.48\, xy^3 z^4 + 3.58\, y^4 z^4 + 0.45\, x^3 z^5 - 0.82\, x^2 yz^5 \\ &- 0.21\, xy^2 z^5 + 4.1\, y^3 z^5 + 1.68\, x^2 z^6 + 0.89\, xyz^6 + 2.35\, y^2 z^6 - 0.38\, xz^7 \\ &+ 1.05\, yz^7 + 0.18\, z^8 \end{aligned}$$

During the course of the classification of minimal immersions of $\mathbb{P}^2$ in $\mathbb{R}^3$, i.e. those which minimise the integral of the squared mean curvature, R. Bryant found a smooth immersion of $\mathbb{P}^2$ in $\mathbb{R}^3$ whose image seems smoothly ambient isotopic to the Boy surface **(Plate 35)**, but the computations allowing us to study a neighborhood of the self-intersection curve have not been carried out:

we use the representation of $\mathbb{P}^2$ as the quotient of the complex projective line $\mathbb{C} \cup \{\infty\}$ by $z \sim -1/\bar{z}$ (representation 8 sec. 1.2) to define the immersion as a map f of $\mathbb{C} \cup \{\infty\}$ into $\mathbb{R}^3$ invariante under $z \to -1/\bar{z}$. Let

$$f(z) = (g_1^2(z) + g_2^2(z) + g_3^2(z))^{-1} \begin{bmatrix} g_1(z) \\ g_2(z) \\ g_3(z) \end{bmatrix}$$

where

$$\begin{aligned} g_1(z) &= -(3/2)\, \mathrm{Im}\,(z\,(1 - z^4)\,(z^6 + \sqrt{5}\,z^3 - 1)^{-1}) \\ g_2(z) &= -(3/2)\, \mathrm{Re}\,(z\,(1 + z^4)\,(z^6 + \sqrt{5}\,z^3 - 1)^{-1}) \\ g_3(z) &= -1/2 + \mathrm{Im}\,((1 + z^6)\,(z^6 + \sqrt{5}\,z^3 - 1)^{-1}) \end{aligned}$$

We will now describe a procedure for constructing a smooth immersion of $\mathbb{P}^2$ in $\mathbb{R}^3$ given by three homogeneous quartic polynomials defined on the sphere; this contradicts the conjecture mentioned by H. Hopf [HO]. Once again the resulting picture ressembles the Boy surface, but the description of a neighborhood of the self-intersection curve seems impracticable. This procedure, which is due to the author, originates from an idea of B. Morin [AP].

Recall that thanks to a remark of D. Hilbert we can assume that the Boy surface has an axis of threefold symmetry. Choose OZ to be the axis of threefold symmetry, and represent the points of the plane XOY by a complex coordinate. We look for the Boy immersion of the form:

$$f(x, y, z) = (g(x, y, z) + jg(y, z, x) + j^2 g(z, x, y), h(x, y, z))$$

where g and h are two homogeneous quartic polynomials on the sphere $S^2 \subseteq \mathbb{R}^3$, the equation of which is $x^2 + y^2 + z^2 = 1$, and $j = \exp(2i\pi/3)$. Denote by p the projection of $\mathbb{R}^3$ onto the plane XOY, and let us define a *critical point* of $p \circ f$ to be any point of the sphere at which the rank of $p \circ f$ is strictly less than 2. The set of critical points will be called the *critical set* of $p \circ f$ and the image of a critical point gives a *critical value*. The set V of critical values of $p \circ f$ is thus the projection by p of the *apparent contour* of the Boy surface seen from the point at infinity on its axis of threefold symmetry.

A study of this apparent contour using Boy's pictures [BOY] shows that it is the image of the circle S^1 under a smooth map whose image has three cusps (Fig. 22). Morin's idea is to interpret $p \circ f$ in a neighborhood of one of these cusps as a perturbation of the *handkerchief folded into quarters* singularity, in other words the mapping (x^2, y^2) in a neighborhood of $(0, 0)$ (Fig. 23). After determination of the projection $p \circ f$ it remains to construct the height function h as a Morse function (sec. 2.1) on the sphere having two maxima six saddles and six minima, in accordance with Boy's indications.

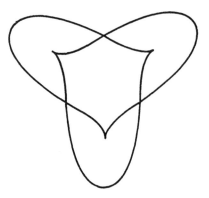

Fig. 22 Apparent contour of the Boy surface.

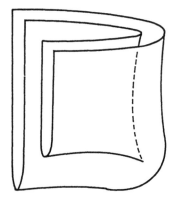

Fig. 23 Perturbation of the folded handkerchief singularity.

2.2 Parametrization of the Boy Surface by 3 Polynomials

By a theorem of A. Haefliger, it is possible to *lift* an arbitrary mapping $\mathbb{P}^2 \xrightarrow{q} \mathbb{R}^2$ to an immersion with target $\mathbb{R}^3$, in other words there is a smooth immersion $\mathbb{P}^2 \to \mathbb{R}^3$ making the following diagram commute

A smooth mapping q from a smooth closed surface M to $\mathbb{R}^2$ will be called *generic* if:

(i) the rank is nowhere equal to zero
(ii) in a neighborhood of any point m at which the rank

is one, a local coordinate system can be chosen in which either $q(x, y) = (x, y^2)$ where $m = (0, 0)$, or $q(x, y) = (x, xy - y^3)$ where $m = (0, 0)$. In the first case the point m is calles a *fold* of q, and is the second case m is called a *pleat* of q. The set of points of M at which the rank of q is strictly less than 2, in other words its critical set, is given locally by $y = 0$ in the first case and $x = 3y^2$ in the second; this shows that the critical set is a smoothly embedded curve without boundary in M. Here is Haefliger's result mentioned above [HA]:

Theorem 1.

A generic mapping $M \xrightarrow{q} \mathbb{R}^2$ can be lifted to a smooth immersion $M \to \mathbb{R}^3$ if and only if the following is true: the number of pleats of q is even on each connected component of the critical set which has an orientable neighborhood, and odd on each component with a nonorientable one (prop. 2 sec. 1.1).

When we have chosen the projection $p \circ f$ we will use this result to examine whether the projection can be lifted. We start with

$$g(x, y, z) = x^2 + k(x, y, z)$$

where the term k is a homogeneous quartic pertubation.

If $k = 0$, the critical set of $p \circ f$ consists of three great circles on the sphere, whose points of intersection form the vertices of a regular octahedron (**Plate 2**). Their image is an equilateral triangle; the critical values situated at the vertices correspond to handkerchief folded into quarters singularities.

To choose this perturbation k, we must investigate how to alter the critical set of the handkerchief folded into quarters singularity. This singularity is not *stable*, in other words a small perturbation can make it disappear: for example, $(x^2 + 2\epsilon y, y^2)$ always has rank greater than or equal to 1, whereas (x^2, y^2) has rank zero at this origin.

Singularity theory (see e.g. [AR] or [MAR]) shows that we can construct a 2-parameter family $k_{u,v}$ such that

$$k_{0,0}(x, y) = (x^2, y^2)$$

and such that near the origin in $\mathbb{R}^4$ the mapping

$$(u, v, x, y) \mapsto (u, v, k_{u, v}(x, y))$$

is, if we make small perturbations, invariant up to changes of coordinate system in the source and in the target. This family is given by

$$k_{u, v}(x, y) = (x^2 + uy, y^2 + vx).$$

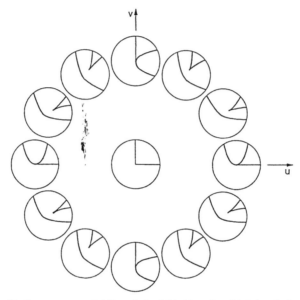

Fig. 24 2-parameter unfolding of the folded handkerchief singularity.

Looking at the set of critical values of $k_{u, v}$ as (u, v) moves around the origin (Fig. 24) shows that, when $uv \neq 0$, it resembles the apparent contour of the Boy surface in a neighborhood of a cusp. We therefore require that the perturbing term k be the product of y and a factor which is nonzero at $(0, 0, 1)$. Similarly, $k(y, z, x)$ must be the product of x and a factor which does not vanish at $(0, 0, 1)$, and so

$$k(x, y, z) = yz(\alpha x^2 + 2\beta xy + \gamma y^2 + 2\delta xz + 2\epsilon yz + \zeta z^2) \qquad \gamma\zeta \neq 0.$$

The critical set of $k_{u, v}$ is the rectangular hyperbola

$$xy = uv/4$$

which is situated in the first and the third quadrants if $u \cdot v$ is positive, and in the second and the fourth quadrants if $u \cdot v$ is negative. Looking at a neighborhood of a vertex of the critical octahedron, e.g. $(0, 0, 1)$, we see that the two cases above are distinguished by the sign of $\gamma\zeta$, since $k(x, y, z) \sim \zeta y$ and $k(y, z, x) \sim \gamma x$. The case $\gamma\zeta > 0$ leads to the critical set of $p \circ f$ having three connected components which does not give the desired apparent contour (Fig. 25) **(Plates 3, 4)**.

2.2 Parametrization of the Boy Surface by 3 Polynomials

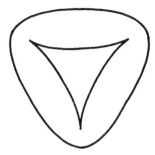

Fig. 25 Apparent contour of the Roman surface.

The simplest case corresponding to $\gamma\zeta<0$ is

$$\gamma=1, \quad \zeta=-1, \quad \alpha=\beta=\delta=\epsilon=0$$

giving

$$g(x,y,z) = x^2 + yz(y^2-z^2) = x^2(x^2+y^2+z^2) + yz(y^2-z^2).$$

The critical set of $p \circ f$ is a connected subset of the sphere **(Plates 5, 6)**, and its image is indeed the required apparent contour (Fig. 22).

To use theorem 1, we need to know whether, when we take the quotient by the antipodal relation, a neighborhood of this curve in $\mathbb{P}^2$ is orientable or nonorientable, that is (by prop. 9 sec. 1.2) whether the critical set is a line or an oval. The following proposition will provide the answer:

Proposition 2.

A smooth embedded circle in $\mathbb{P}^2$ which intersects a given smooth line d transversally, is an oval or a line depending on whether the number of points of intersection is even or odd. The number of points of intersection of two ovals meeting transversally is even.

We say that the images of two smooth embeddings of the circle intersect *transversally*, if at each point of intersection the tangent vectors to the images are linearly independent. The proof of prop. 2 if left to the reader; the idea is to deform by a homotopy to the case of a standard oval and a projective line, and to note that the parity of the number of points of intersection remains constant.

In the case without the perturbing term k, the critical set of $p \circ f$ is given by three great circles

$$x=0 \quad y=0 \quad z=0$$

which intersect the great circle $x+y+z=0$ transversally at the six points $(0, 1, -1)$ $(0, -1, 1)$ $(1, 0, -1)$ $(-1, 0, 1)$ $(1, -1, 0)$ $(-1, 1, 0)$. If we make a slight perturbation of this critical set the number of points of intersection remains constant, and passing to the quotient, we obtain three points of intersection with the line which is the projection to $\mathbb{P}^2$ of the great circle $x+y+z=0$. Since this critical set is connected, the preceding proposition shows that it gives a line in the quotient space $\mathbb{P}^2$; this line has a nonorientable neighborhood.

The set of critical values has three cusps coming from the three pleats on the critical set in $\mathbb{P}^2$. Theorem 1 shows that we can lift the map of $\mathbb{P}^2$ to $\mathbb{R}^2$ resulting from $p \circ f$ to a smooth immersion with target $\mathbb{R}^3$. We will now determine the height function h as a homogeneous *alternating* quartic polynomial, i.e. one which is invariant under cyclic permutation of the variables; we insist on this so as to respect the threefold symmetry of the Boy surface. The choice of the function

$$h(x, y, z) = (x + y + z)^4/4 + (x + y + z)(y - x)(z - y)(x - z)$$

will be justified by the following result:

Lemma 3.

Let G denote the group of those isometries of $\mathbb{R}^3$ generated by the central inversion about the point 0, and by the rotation through $2\pi/3$ about the axis $x = y = z$; let a be a real root of $70 a^3 - 186 a^2 + 171 a - 54 = 0$. Then h is a Morse function on the sphere $S^2 \subseteq \mathbb{R}^3$ with two maxima at the points of the G-orbit of $(1/\sqrt{3}, 1/\sqrt{3}, 1/\sqrt{3})$, six saddles given by the G-orbit of $(\sqrt{2}/\sqrt{3}, -1/\sqrt{6}, -1/\sqrt{6})$, and six minima at the points of the G-orbit of $(-((1-a)/3)^{1/2}, -(a/2)^{1/2} - ((1-a)/3)^{1/2}, (a/2)^{1/2} - ((1-a)/3)^{1/2})$.

The proof of this lemma is a simple numerical calculation. We have now found the desired mapping $S^2 \xrightarrow{f} \mathbb{R}^3$, but it remains to prove that the rank is indeed 2 at every point (**Plates 36–38**).

Proposition 4.

The map

$$f(x, y, z) = \begin{bmatrix} ((2x^2 - y^2 - z^2)(x^2 + y^2 + z^2) + 2yz(y^2 - z^2) + zx(x^2 - z^2) + xy(y^2 - x^2))/2 \\ ((y^2 - z^2)(x^2 + y^2 + z^2) + zx(z^2 - x^2) + xy(y^2 - x^2)) \cdot \sqrt{3}/2 \\ (x + y + z)((x + y + z)^3/4 + (y - x)(z - y)(x - z)) \end{bmatrix}$$

defined on the sphere yields, on factorizing by the antipodal action, a smooth immersion of $\mathbb{P}^2$ in $\mathbb{R}^3$ whose image is contained in, but not in equal to, the real zeroset of an irreducible polynomial of degree sixteen.

The details of the proof can be found in [AP]. We remark that the difficulty is due to the fact that the mapping f fails to have a maximal rank if we allow x, y, z to take complex values; this prevents us from using the *resultant*. Furthermore the irreducible polynomial which vanishes on the image of f also vanishes at the point $(0, 0, -3)$ which is not part of this image. The preimages of the above point are found by solving the equations

$$x + y + z = \sqrt{2} \pm i, \quad xy + yz + zx = \pm i\sqrt{2}, \quad xyz = -(3\sqrt{2} \pm i)/12$$
$$(y - x)(z - y)(x - z) = -(11\sqrt{2} \pm 2i)/3.$$

In the next sections we shall construct a polynomial which will be free from any such blemishes since its set of real zeroes will be precisely the Boy surface. It is legitimate to ask whether it might be possible to immerse $\mathbb{P}^2$ smoothly in $\mathbb{R}^3$, using a mapping from S^2 to $\mathbb{R}^3$ whose components are three homogeneous quadratic polynomials. We can reply to this question in the negative:

Proposition 5.

There is no smooth immersion of $\mathbb{P}^2$ in $\mathbb{R}^3$ given by three quadratic forms on the sphere.

Proof.

By prop. 7 sec. 1.1, such a mapping cannot be injective. Let d be a projective line in $\mathbb{P}^2$ joining two points which have the same image. Since line d is compact, its image is also compact; in addition the image is the projection of a conic, and it is therefore a line segment. The mapping fails to have maximal rank at each point whose image is one of the two ends of the segment.   q.e.d.

2.3 How to Eliminate Whitney Umbrellas

We have seen in section 1.3 that the presence of singularities introduces certain defects, both algebraic and geometric, in the representation of $\mathbb{P}^2$ by means of Steiner surfaces. We now propose to describe these singularities and to give a procedure for eliminating them.

If U denotes an open subset of $\mathbb{R}^m$, we write $C^\infty(U, \mathbb{R}^n)$ for the vector space of smooth mappings from U to $\mathbb{R}^n$, equipped with the topology of uniform convergence on each compact subset for all the partial derivatives (including that of order 0). We say that the map $f \in C^\infty(U, \mathbb{R}^n)$ is *stable at* $a \in U$ if there exists a neighborhood N of f in $C^\infty(U, \mathbb{R}^n)$ with the following property: for every g in N there should exist smooth diffeomorphisms $U' \xrightarrow{h} U''$ and $V' \xrightarrow{k} V''$, where U' is an open neighborhood of a contained in U, U'' is an open subset of U, and V' and V'' are open neighborhoods of $f(U')$ and $g(U'')$ respectively, such that there is a commutative square

$$\begin{array}{ccc} U' & \xrightarrow{f} & V' \\ h \downarrow & & \downarrow k \\ U'' & \xrightarrow{g} & V'' \end{array}$$

Example 1.

If the rank of f is maximal at $a \in U$, then f is stable at a. This is a consequence of the *rank theorem* [SP] which states that:

If $M \xrightarrow{f} N$ is a smooth mapping of smooth manifolds whose rank is equal to a constant r in a neighborhood of a, then there are local coordinate systems h and k at a and $f(a)$ such that $kfh^{-1}(x_1, ..., x_m) = (y_1, ..., y_r, 0, ..., 0)$.

Example 2.

If $U \xrightarrow{f} \mathbb{R}$ is a smooth function on an open subset of $\mathbb{R}^m$, we define a *nondegenerate critical point of index* r by extending in an obvious manner the definitions of sec. 2.1. If a function f has a nondegenerate critical point at a in U, then f is stable at a. This is a corollary of the Morse lemma [HI] which is as follows:

Let f be a smooth real-valued function on a smooth manifold M which has a non-degenerate critical point at a of index r. Then there is a local coordinate system h at a such that:

$$fh^{-1}(x_1, \ldots, x_m) = f(a) - \sum_{i=1}^{r} x_i^2 + \sum_{i=r+1}^{m} x_i^2 .$$

Let U be an open subset of $\mathbb{R}^m$. We define a *singular point* of a smooth mapping $U \xrightarrow{f} \mathbb{R}^n$ to be any point at which the rank of f fails to be maximal, i.e. where the rank is strictly less than the smaller of m and n. If a is a singular point of f, the *singularity at a* is the equivalence class of f modulo the following equivalence relation: we identify two smooth maps $U \xrightarrow{f} \mathbb{R}^n$ and $U' \xrightarrow{g} \mathbb{R}^n$ if there exist open neighborhoods V and V' of a contained in U and U', open neighborhoods W and W' of $f(a)$ containing $f(V)$ and $g(V')$, and smooth diffeomorphisms $V \xrightarrow{h} V'$ and $W \xrightarrow{k} W'$ such that we have a commutative diagram:

$$\begin{array}{ccc} V & \xrightarrow{f} & W \\ h \downarrow & & \downarrow k \\ V' & \xrightarrow{g} & W' \end{array}$$

Example 3.

The fold and the pleat given by (x, y^2) and $(x, xy - y^3)$ are the only stable singularities of mappings from $\mathbb{R}^2$ to $\mathbb{R}^2$; this theorem is due to H. Whitney [MAR]. A mapping from a smooth closed surface to $\mathbb{R}^2$ which is generic in the sense of sec. 2.2 is thus stable at every point.

Example 4.

As we indicated in sec. 2.2, the folded handkerchief singularity given by the mapping

$$f(x, y) = (x^2, y^2)$$

from $\mathbb{R}^2$ to $\mathbb{R}^2$ is not stable at the origin $(0, 0)$. Nevertheless, we have seen that this singularity can be unfolded to obtain the mapping of $\mathbb{R}^4$ into $\mathbb{R}^4$ given by

$$g(u, v, x, y) = (u, v, x^2 + uy, y^2 + vx)$$

This mapping has a singularity at the origin which we call a *hyperbolic umbilic* (Fig. 26). This map is stable [MAR].

It was shown in 1944 by Hassler Whitney [WH2] that the only stable singularity of mappings from $\mathbb{R}^2$ to $\mathbb{R}^3$ is that which is called the *Whitney umbrella*. Using an appropriate local coordinate system at the origin, it can be written

$$f(x, y) = (x, xy, y^2) .$$

The image of this parametrization is part of the ruled cubic surface with equation

$$Y^2 = X^2 Z .$$

2.3 How to Eliminate Whitney Umbrellas

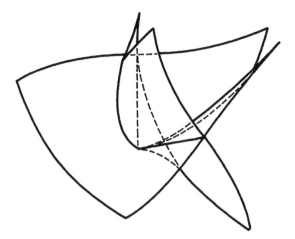

Fig. 26 Section by the hyperplane $u = v$ of the set of critical values of the hyperbolic umbilic $(u, v, x^2 + uy, y^2 + ux)$.

But, once again, we encounter the phenomenon which we have already met at the end of sec. 1.3; the cubic surface contains the half-line

$$Y = X = 0 \qquad Z < 0$$

which is not in the image of the parametrization **(Plate 23)**. This half-line is called the *handle of the Whitney umbrella*.

At each point of the other half of this line, except the origin, the surface has two tangent planes whose intersection is the line containing the handle, and whose angle of intersection tends to zero as the point approaches the origin $(0, 0, 0)$. We define the *contingent* at a of a subset E of $\mathbb{R}^n$, to be the set of limiting positions of half-lines with origin a, passing through the point x in E as x tends to a. The contingent at $(0, 0, 0)$ of the image of f is the half-plane $Y = 0$ and $Z \geqslant 0$, which means that there can not be a tangent plane.

This singularity, the Whitney umbrella, occurs at the two points $(0, 0, 0)$ and $(0, 0, 1)$ on the cross-cap of Steiner (sec. 1.3) **(Plates 18–20)**, and at the six points $(0, 0, 1/2)$ $(0, 0, -1/2)$ $(0, 1/2, 0)$ $(0, -1/2, 0)$ $(1/2, 0, 0)$ $(-1/2, 0, 0)$ on Steiner's roman surface **(Plates 21, 22)**.

We observe that the properties of stability and of the singularity at a point a, are not affected if we replace the smooth mapping by another mapping which coincides with the first one in some neighborhood of a. If we regard two smooth $\mathbb{R}^n$-valued mappings which coincide in a neighborhood of a as equivalent, the equivalence classes are called *germs at a with values in $\mathbb{R}^n$*.

We now mention a very powerful tool for studying singularities: the Malgrange-Mather preparation theorem. The following version is taken from [MAR]; denote by E_p the algebra of germs of real-valued functions defined on $\mathbb{R}^p$ which vanish at the origin; the usual arithmetic operations on $\mathbb{R}$ give E_p an algebra structure. Given a map germ f

from $\mathbb{R}^p$ to $\mathbb{R}^n$ which vanishes at the origin, we can take any module M over E_n and provide it with a module structure over E_p by the following action

$$g \cdot m = g \circ f \cdot m$$

Here g is an element of E_p and m is an element of M. We now state the Malgrange-Mather theorem

Theorem 1.

M is generated as module over E_p by $m_1, \ldots, m_k$ if and only if every element m of M can be written in the form

$$m = a_1 m_1 + \ldots + a_k m_k + f_1 m_1' + \ldots + f_p m_p'$$

where the a_i are real numbers, the m_i' are elements of M, and the f_i are the components of the germ f.

Thanks to this result, we can easily identify Whitney umbrellas:

Proposition 2.

A map germ f from $\mathbb{R}^2$ to $\mathbb{R}^3$, whose components will be denoted by f_1, f_2, f_3, has a Whitney umbrella singularity at the origin if and only if the two following conditions are satisfied:

(i) at least one entry in the jacobian matrix of f is nonzero at the origin; by permuting the coordinate axis in source and target, we shall assume $\partial_x f_1 (0,0) \neq 0$

(ii) the determinants

$$D_2 = \begin{vmatrix} \partial_x f_1 & \partial_y f_1 \\ \partial_x f_2 & \partial_y f_2 \end{vmatrix} \qquad D_3 = \begin{vmatrix} \partial_x f_1 & \partial_y f_1 \\ \partial_x f_3 & \partial_y f_3 \end{vmatrix}$$

vanish at the origin, but the jacobian determinant of (D_2, D_3) is nonzero at the origin.

Proof.

It is clear that the umbrella singularity satisfies these conditions. To show the converse, we will prove that by making local coordinate changes in source and target, we can write f in the normal form given above in the definition of the Whitney umbrella.

By making a change of local coordinates in the target, we may suppose that $f(0,0) = 0$, and the hypothesis $\partial_x f_1 (0,0) \neq 0$ allows us to choose f_1 to be the first local coordinate in the source:

$$f_1(x, y) = x.$$

The condition $D_2 = D_3 = 0$ means that $\partial_y f_2 (0,0) = \partial_y f_3 (0,0) = 0$. Morever, since the jacobian determinant of D_2 and D_3 is nonzero at the origin, we can make a local coordinate change in the target which leaves f_1 unaltered, and allows us to assume that

$$\partial_{y^2} f_2 (0,0) = 0 \qquad \partial_{y^2} f_3 (0,0) \neq 0.$$

2.3 How to Eliminate Whitney Umbrellas

Using a Taylor expansion with integral remainder term, we see that

$$f_3(x, y) = x g_3(x, y) + y^2 h_3(x, y)$$

where g_3 and h_3 are real-valued germs at the origin in $\mathbb{R}^2$, and h_3 is invertible. We may therefore write

$$x = f_1 \quad y^2 = f_3/h_3 - f_1 \cdot g_3/h_3$$

and thus, once again using Taylor's formula with integral remainder, any real-valued germ k at the origin in $\mathbb{R}^2$ can be written

$$k(x, y) = a + by + xc(x, y) + y^2 d(x, y) = a + by + f_1 \cdot (c - g_3 d/h_3) + f_3 \cdot d/h_3$$

where a and b are real numbers, and c and d are function germs at the origin in $\mathbb{R}^2$. Applying theorem 1 to the module of germs $k(x, y)$, we find that all these germs, in particular y^2, can be expressed in the form

$$y^2 = \varphi(f_1, f_3) + y \cdot \psi(f_1, f_3).$$

If we make the local change of coordinates in the source given by

$$x' = x$$
$$y' = y - \psi(f_1, f_3)/2$$

whose jacobian matrix at the origin is the identity matrix, then the above properties of f_2 and f_3 at the origin are preserved. Finally, we may suppose that $\psi(f_1, f_3) = 0$ and hence write

$$y^2 = \varphi(f_1, f_3).$$

We have seen that

$$f_3(0, y) = y^2 h_3(0, y)$$

where h_3 is a germ at the origin with $h_3(0, 0) \neq 0$. Denoting by X, Y, Z the local coordinates in the target, we deduce that

$$\varphi(0, Z) = Z/h_3(0, 0) + Z^2 \chi(Z).$$

This shows it is legitimate to take as third local coordinate in the target $\varphi(X, Z)$ instead of Z, which amounts to assuming that

$$f_3(x, y) = y^2.$$

Applying theorem 1 in the same way as above, we find that any real-valued germ at the origin in $\mathbb{R}^2$, in particular f_2, can be written in the form

$$f_2(x, y) = g_2(x, y^2) + y h_2(x, y^2).$$

By the local coordinate change in the target given by

$$X' = X$$
$$Y' = Y - g_2(X, Z)$$
$$Z' = Z$$

we see that we can put $g_2(x, y) = 0$. The hypothesis that the jacobian determinant of (D_2, D_3) is nonzero at the origin means that

$$\partial_{xy} f_2(0, 0) \neq 0.$$

Looking at the germ $h_2(X, Z)$, we have

$$\partial_{xy} f_2(x, y) = \partial_x h_2(x, y^2) + 2y^2 \partial_{xz} h_2(x, y^2)$$

which proves that $\partial_x h_2(0, 0)$ is nonzero. We therefore make a local coordinate change in the source

$$x' = h_2(x, y^2)$$
$$y' = y$$

this will affect f_1, and so we get it back by making the final change of local coordinates in the target

$$X' = h_2(X, Z) \quad Y' = Y \quad Z' = Z$$

which like its predecessor is a legitimate transformation. To conclude we have now obtained the desired normal form:

$$X' = x' \quad Y' = x'y' \quad Z' = y'^2 \qquad \text{q.e.d.}$$

Example 5.

Consider the singularity at the origin defined by the map germ

$$f(x, y) = (x, xy, y^3)$$

We find $\partial_x f_1 = 1$, $D_2(0, 0) = D_3(0, 0) = 0$ but the determinant

$$\begin{vmatrix} \partial_x D_2 & \partial_y D_2 \\ \partial_x D_3 & \partial_y D_3 \end{vmatrix} = 6y$$

vanishes at $(0, 0)$. This singularity is not stable; for by a small perturbation we can get the map $(x, xy, y^3 + \epsilon y^2)$ whose only singularities are Whitney umbrellas (Fig. 27).

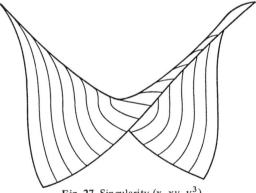

Fig. 27 Singularity (x, xy, y^3).

2.3 How to Eliminate Whitney Umbrellas

Example 6.

The parametrization of Steiner's cross-cap (sec. 1.3) in a neighborhood of the point $(0, 0, 1)$ is given by

$$f(x, y) = (yz, 2xy, x^2 - y^2)$$

where

$$z = (1 - x^2 - y^2)^{1/2}$$

The jacobian matrix of f is

$$\begin{bmatrix} -xy/z & z - y^2/z \\ 2y & 2x \\ 2x & -2y \end{bmatrix}$$

We have $\partial_y f_1 (0, 0) = 1$, $D_2 = -2yz + 2y^3/z - 2x^2 yz$ and $D_3 = 4xy^2/z - 2xz$; so we can check that $D_2 (0, 0) = D_3 (0, 0) = 0$ and that the jacobian determinant of (D_2, D_3) is equal to -4 at the origin. Therefore we have a Whitney umbrella.

Starting from a slight modification of the real Veronese map (sec. 1.3), define an embedding of $\mathbb{P}^2$ in $\mathbb{R}^6$ by factorizing the smooth embedding of S^2 in $\mathbb{R}^6$ given by

$$f(x, y, z) = (x^2, y^2, z^2, \sqrt{2}yz, \sqrt{2}zx, \sqrt{2}xy).$$

We can construct an embedding of $\mathbb{P}^2$ in $\mathbb{R}^4$ by observing that the image of f is contained in the 4-dimensional sphere S^4 given by

$$X_1^2 + X_2^2 + X_3^2 + X_4^2 + X_5^2 + X_6^2 = 1$$

$$X_1 + X_2 + X_3 = 1.$$

We now remark that the image of f is not the whole of S^4, for instance it does not include the point

$$(1/2, (1 - \sqrt{5})/4, (1 + \sqrt{5})/4, 0, 0, 0).$$

So, by projecting stereographically from this point, we obtain a smooth embedding of $\mathbb{P}^2$ in $\mathbb{R}^4$.

Another result of Whitney [WH1] proves that we can not project a smooth embedding of $\mathbb{P}^2$ in $\mathbb{R}^4$ to get a smooth immersion of $\mathbb{P}^2$ in $\mathbb{R}^3$. However, we can arrange that the only singularities that occur will be Whitney umbrellas, since

Proposition 3.

If f is a smooth immersion of the closed surface M in $\mathbb{R}^4$, then there exists a projection p from $\mathbb{R}^4$ to $\mathbb{R}^3$ such that the only singularities of the composite $p \circ f$ are Whitney umbrellas.

Proof.

Consider a system of local coordinates on M

$$U \xrightarrow{h} hU \subseteq \mathbb{R}^2$$

Using this system we can define the jacobian matrix J of $f \cdot h^{-1}$. The matrix J induces a smooth map from $hU \times \mathbb{R}^2$ to $\mathbb{R}^4$, given by

$$\varphi(x, y, a, b) = J(x, y) \begin{pmatrix} a \\ b \end{pmatrix}.$$

The set of critical values of φ, that is the set of images of points at which the rank is not equal to 4, is of measure zero in $\mathbb{R}^4$ according to a theorem of Sard [HI]. The union of these sets for each of the local coordinate systems of a finite atlas of M is a subset of $\mathbb{R}^4$ of measure zero which is independent of the choice of atlas.

We can therefore choose the projection mapping p such that its kernel is not a subset of this set of measure zero. Let us consider a point of M at which the rank of $p \circ f$ is less than 2, in other words where the rank is 1. We may suppose that

$$p(X_1, X_2, X_3, X_4) = (X_1, X_2, X_3) \quad \partial_x g(0, 0) \notin \ker p \quad \partial_y g(0, 0) \in \ker p$$

where $g = f_0 h^{-1}$, and the point in question has local coordinates $(0, 0)$. We can choose local coordinates in which

$$g_1(x, y) = x \quad \partial_x g_2(0, 0) = 0 \quad \partial_x g_3(0, 0) = 0 \quad g_4(x, y) = y.$$

Because of our choice of the kernel of the projection p, the mapping φ defined above has rank 4 at $(0, 0, 0, 1)$, which means

$$\begin{vmatrix} 0 & 0 & 1 & 0 \\ \partial_{xy} g_2 & \partial_{y^2} g_2 & 0 & 0 \\ \partial_{xy} g_3 & \partial_{y^2} g_3 & 0 & 0 \\ 0 & 0 & 0 & 1 \end{vmatrix} = \begin{vmatrix} \partial_{xy} g_2 & \partial_{y^2} g_2 \\ \partial_{xy} g_3 & \partial_{y^2} g_3 \end{vmatrix} \neq 0$$

This gives precisely the conditions required in proposition 2 for a Whitney umbrella.

q.e.d.

Because he could not see how to eliminate the Whitney umbrellas which are present on the Steiner surfaces, Hilbert believed that it was impossible to immerse the projective plane $\mathbb{P}^2$ in $\mathbb{R}^3$; this was before the result of Boy.

We now know that these singularities are born and annihilated in pairs in a process which is generic in 1-parameter families of mappings from $\mathbb{R}^2$ to $\mathbb{R}^3$. We note that the Steiner surfaces have an even number of Whitney umbrellas (there are two on the crosscap and six on the Roman surface). Even the ruled cubic surface which defines the standard model of an umbrella **(Plate 23)** has a second singularity, at infinity. We can see this by applying the homography

$$\begin{bmatrix} X' \\ Y' \\ Z' \\ T' \end{bmatrix} = \begin{bmatrix} 1 & 0 & 0 & 0 \\ 0 & 1 & 0 & 0 \\ 0 & 0 & 1 & 0 \\ 0 & 0 & 1 & 1 \end{bmatrix} \begin{bmatrix} X \\ Y \\ Z \\ T \end{bmatrix}$$

which gives the *Plücker conoid* **(Plate 24)** with equation

$$Y'^2 = (X'^2 + Y'^2) Z'$$

2.3 How to Eliminate Whitney Umbrellas

which has umbrellas at (0, 0, 0) and (0, 0, 1). This surface can be described geometrically as the union of the common perpendiculars to OZ' and those lines which meet both of the lines $Y' = Z' = 0$ and $X' = Z' - 1 = 0$. Some of the many interesting properties of this surface where given in [DA] and more recently in [FI].

In this context, we point out what happens if we perform an inversion with respect to a point and the axis of symmetry OZ' which is not one of the Whitney umbrellas, say $(0, 0, -1)$, which interchanges the two umbrellas, in other words, the ratio of the inversion is 2. We obtain a quartic surface generated by circles tangent at the point $(0, 0, -1)$ to the plane $Z' = -1$; the equation of the surface is

$$(X'^2 + Y'^2)^2 + Y'^2 (X'^2 + Y'^2 + (Z' + 1)^2) + (Z'^2 - 1)(X'^2 + Y'^2) = 0 .$$

Even though this algebraic surface has the same degree of the Steiner surfaces, and furthermore its real part is smoothly ambient isotopic (sec. 2.2) to the cross-cap, this does not imply that this surface is in fact the Steiner cross-cap (Fig. 28).

Fig. 28 Quartic surface obtained by inversion of the Plücker conoid.

To identify these surfaces would be to overlook the fact that the Steiner surface has a triple point which is the intersection of three noncoplanar double lines, only one of which is visible in $\mathbb{R}^3$. These lines force the generic plane section to be a rational quartic whereas for the inverse of the Plücker conoid a generic plane section is of genus 2.

To eliminate a pair of Whitney umbrella singularities of a smooth map f from $\mathbb{R}^2$ to $\mathbb{R}^3$, we have to deform it by means of a parameter t, which amounts to considering a one-parameter family f_t of mappings from $\mathbb{R}^2$ to $\mathbb{R}^3$.

For the value of the parameter t which corresponds to the the birth or death of the pair of umbrellas, say $t = 0$, the singularity which appears at this point must necessarily be unstable, since the only stable singularity of mappings from $\mathbb{R}^2$ to $\mathbb{R}^3$ is the Whitney umbrella. To investigate this singularity we must *unfold* it, in other words we consider a smooth mapping F from $\mathbb{R}^3$ to $\mathbb{R}^4$ such that

$$F(t, x, y) = (t, f_{t1}(x, y), f_{t2}(x, y), f_{t3}(x, y)) .$$

Unfortunately, the concept of stability which we introduced above will not be useful here, since a result of B. Morin [MO1] tells us that the only stable singularity of map germs from $\mathbb{R}^3$ to $\mathbb{R}^4$ is given, using appropriate local coordinate systems, by the germ at the origin of

$$F(t, x, y) = (t, x, y^2, xy)$$

which is called the *suspended umbrella*.

This germ does not display the birth or death of umbrellas when the parameter t passes from negative to positive values; rather for all value of t the germ

$$(x, y) \mapsto F(t, x, y)$$

is a Whitney umbrella. The *curve of umbrellas* of the germ F, that is to say the set of points at which the rank of F is not maximal, has equation $x = y = 0$.

If, by a suitable local coordinate change at the origin in $\mathbb{R}^3$, we can arrange for the curve of umbrellas to be tangent to the plane $t = 0$ without crossing it, then we will indeed have brought about the birth or death of two umbrellas as the parameter t passes through the value 0 (Fig. 29).

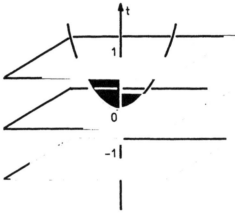

Fig. 29 Curve of umbrellas of a germ of suspended umbrella type. We can choose a deformation parameter t such that this curve is tangent to the plane t = 0.

We will use the term *confluence of Whitney umbrellas* to denote any deformation f_t induced by a germ of the type of the suspended umbrella F for which the contact at the origin between the curve of umbrellas and the plane $t = 0$ is precisely of second order; furthermore, we require that the restriction of F to this curve is regular at $(0, 0, 0)$. We now state the following proposition [WA]:

Proposition 4.

There are exactly two deformations of the type confluence of Whitney umbrellas given in appropriate local coordinates at the origin by

$$f_t(x, y) = (x, y^2, x^2 y + y^3 - ty)$$
$$g_t(x, y) = (x, y^2, x^2 y - y^3 - ty)$$

2.3 How to Eliminate Whitney Umbrellas

The first type is called the *elliptic confluence* of umbrellas and the second the *hyperbolic confluence*.

Let us denote the jacobian determinant of $(f, ..., g)$ with respect to the variables $(x, ..., y)$ by

$$J(f, ..., g; x, ..., y)$$

In practice, we can characterize the above deformations by

Proposition 5.

The deformation $f_t(x, y)$ induced by the germ F from $\mathbb{R}^3$ to $\mathbb{R}^4$ at the origin, given by

$$F(t, x, y) = (t, F_1(t, x, y), F_2(t, x, y), F_3(t, x, y))$$

is a confluence of Whitney umbrellas if and only if the following conditions are satisfied:

(i) the rank of F at the origin is equal to two; in other words, f_0 has rank one at the origin in $\mathbb{R}^2$. We shall suppose that $\partial_x F_1 \neq 0$ and $\partial_y F_1 \neq 0$ at $(0, 0, 0)$, and that the two jacobian determinants

$$D_2 = J(F_1, F_2; x, y) \qquad D_3 = J(F_1, F_3; x, y)$$

vanish at $(0, 0, 0)$: we can always for this to be so permute the coordinates in the target.

(ii) the equations $D_2 = 0$ and $D_3 = 0$ are independent in a neighborhood of the origin, and locally they define a curve, called the *curve of umbrellas*, which is tangent to the plane $t = 0$ at $(0, 0, 0)$. In other words

$$J(D_2, D_3; x, y)(0, 0, 0) = 0$$

and in addition (modulo a permutation of the coordinates in the source)

$$J(D_2, D_3; t, y)(0, 0, 0) \neq 0.$$

(iii) the contact between the curve of umbrellas and the plane $t = 0$ is precisely second order; if, for example, $\partial_y D_2 \neq 0$ at $(0, 0, 0)$ we require that

$$J(J(D_2, D_3; x, y), D_2; x, y) \neq 0$$

(iv) the restriction of F to the curve of umbrellas is regular at the origin; this means that at least one of the jacobian determinants of (D_2, D_3, F_1) (D_2, D_3, F_2) (D_2, D_3, F_3) does not vanish at the origin. If, for instance, $\partial_y D_2 \neq 0$ at $(0, 0, 0)$ then we require that at least one of the determinants

$$J(F_1, D_2; x, y) \qquad J(F_2, D_2; x, y) \qquad J(F_3, D_2; x, y)$$

should be nonzero at $(0, 0, 0)$.

This condition is precisely the translation of the definition given above of a confluence of umbrellas into the language of jacobian determinants. To distinguish between hyperbolic and elliptic confluence, we check whether there is a neighborhood of the origin in $\mathbb{R}^2$, independent of the parameter t, such that f_t restricts to an embedding of this neighborhood as t passes through the value 0. Let us examine the canonical forms of these confluences as given in proposition 4.

Elliptic Confluence

The deformation is given by

$$f_t(x, y) = (x, y^2, x^2y + y^3 - ty)$$

We remark that the image of f_t is a subset of the real part of the quintic surface

$$Z^2 = Y(X^2 + Y - t)^2$$

whose set of singular points is the parabola

$$Z = 0 \quad Y = t - X^2.$$

For $t > 0$ the image of f_t includes only that part of the parabola for which Y is greater than or equal to zero. This is because of the presence of two Whitney umbrella singularities at the points $(\sqrt{t}, 0)$ and $(-\sqrt{t}, 0)$ whose images are $(\sqrt{t}, 0, 0)$ and $(-\sqrt{t}, 0, 0)$; the handles are the two arcs of the parabola corresponding to negative values of Y **(Plate 25)**.

When $t = 0$, the singular parabola becomes tangent to the plane $Z = 0$, and the two umbrellas merge to form the confluence singularity at $(0, 0)$. The mapping f_0 is topologically an embedding, but it is not smooth at $(0, 0)$ **(Plate 26)**.

For negative t, the parabola does not meet the image of f_t; the two umbrellas have disappeared and f_t is a smooth embedding **(Plate 27)**.

Hyperbolic Confluence

The deformation can be written as

$$g_t(x, y) = (x, y^2, x^2y - y^3 - ty).$$

The image of g_t is contained in the real part of the quintic surface

$$Z^2 = Y(X^2 - Y - t)^2$$

whose set of singular points is the parabola

$$Z = 0 \quad Y = X^2 - t.$$

For positive t, the image of g_t includes only those points of the parabola corresponding to values of Y which are positive or zero. This indicates the presence of umbrellas at $(\sqrt{t}, 0, 0)$ and $(-\sqrt{t}, 0, 0)$ whose handles are formed by that part of the parabola with $Y < 0$ **(Plate 28)**.

When t is zero, the singular parabola becomes tangent to the plane $Z = 0$ and the two umbrellas come together into the confluence singularity at $(0, 0)$. In contrast to the elliptic case, g_0 is not a topological embedding since there are still double points in the neighborhood of the origin, given by the pairs (x, x) and $(x, -x)$ which have the same image on the singular parabola **(Plate 29)**.

For negative t, the two umbrellas have disappeared and g_t is now a smooth immersion having the singular parabola as double curve; this parabola is wholly contained in the image of g_t **(Plate 30)**.

2.3 How to Eliminate Whitney Umbrellas

Note that at every point of the singular parabola which is not an umbrella point the quintic surface has two distinct tangent planes, which are real if the point is in the image of the deformation; otherwise they are complex conjugates. The two tangent planes coincide at an umbrella.

Proposition 3 shows how to construct a smooth map f from a closed surface M to $\mathbb{R}^3$ whose only singularities are Whitney umbrellas. The umbrellas are at the end points of curves in the self-intersection set of f, and this leads to Whitney's result that the number of umbrellas is even.

To eliminate a pair, we choose a smooth path on M from one umbrella to the other; this is a smooth map from the unit interval I to M. Such a path corresponds, in the canonical models for the confluences described above, to

$$u \mapsto ((u-1)\sqrt{t} + u\sqrt{t}, 0).$$

We then perform a confluence along the selected path.

The most striking example is to produce a confluence on Steiner's cross cap (sec. 1.3) along an S-shaped curve joining the umbrellas at $(0, 0, 0)$ and $(0, 0, 1)$ and meeting the self-intersection interval

$$X = Y = 0 \qquad 0 < Z < 1.$$

The result of this operation (Fig. 30) is remarkably similar to the first version of the surface discovered by Boy (see [BOY] or [FI]).

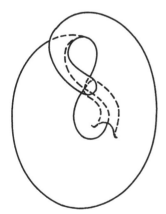

Fig. 30 Surface obtained from the Steiner cross-cap by hyperbolic confluence of two umbrellas along a curve in S

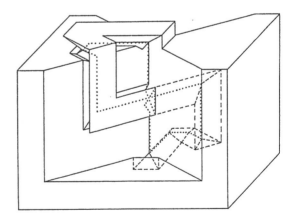

Combinatorical model of the left surface

In the next section we shall construct a symmetric version of the Boy surface by simultaneously eliminating, through confluence, the three pairs of umbrellas on Steiner's Roman surface, in such a way that throughout this process we preserve an axis of threefold symmetry.

2.4 A Boy Surface of Degree Six

The idea of constructing the Boy surface by deforming the Steiner Roman surface so as to eliminate the six Whitney umbrellas in pairs is due to B. Morin, who noticed a certain number of similarities in the geometry of the two surfaces.

By the observation of Hilbert mentioned above (sec. 2.2), we can take a Boy surface having an axis of threefold symmetry. This axis meets the surface at its triple point and also at a second point which we call the *pole of the Boy surface*. The Roman surface has the symmetry of the regular tetrahedron; it has four axis of threefold symmetry which intersect at the triple point and which meet the surface at four other points called the *poles of the Roman surface* **(Plates 21, 22)**.

The Roman surface is generated by ellipses passing through one of its poles, and the empirical parametrization described in [PE] (sec. 2.2) shows that we can attempt to generate the Boy surface by ellipses passing through its pole **(Plates 31, 32)**.

In the case of the Roman surface, these ellipses lie in planes whose envelope is the apparent contour seen from the pole; this apparent contour is a cone on a hypocycloid with three cusps (Fig. 31). B. Morin pointed out that the apparent contour of the Boy surface seen from its pole is a cone on a closed curve with no double point, three cusps, and of course an axis of threefold symmetry. It is natural to try to make it into a hypocycloid with three cusps.

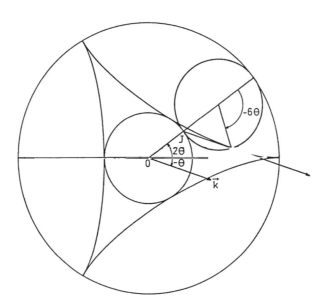

Fig. 31 A mechanism for tracing the hypocycloid with three cusps.

2.4 A Boy Surface of Degree Six

Let us examine in greater detail the generation of the Roman surface by ellipses passing through a pole. We shall consider the Roman surface as excluding the handles of the six umbrellas (sec. 2.3); in other words we study the image in of the projective plane $\mathbb{P}^2$ under the mapping of section 1.3:

$$(yz, zx, xy, x^2 + y^2 + z^2).$$

Since we want to construct an immersion of $\mathbb{P}^2$ in $\mathbb{R}^3$, we fix the plane $T = 0$ to be the plane at infinity, and we identify its complement with $\mathbb{R}^3$ with coordinates X, Y, Z; the plane at infinity $T = 0$ does not intersect the image of the above parametrization.

The Roman surface appears as the image of the sphere S^2 with equation $x^2 + y^2 + z^2 = 1$ in $\mathbb{R}^3$ under the parametrization

$$(yz, zx, xy)$$

which passes to $\mathbb{P}^2$ after factoring by the action of the antipodal map.

The two preimages of the pole of the Roman surface are two antipodal points on the sphere: the poles. The ellipses on the Roman surface passing through the pole are the images of the great circles of the sphere passing through its two poles, or rather of the meridians which are halves of these great circles, since the circles are double coverings of the ellipses.

Passing to the quotient under the antipodal map, these meridians become projective lines in $\mathbb{P}^2$. Thus the ellipses on the Roman surface passing through the pole are the images of the pencil of projective lines in $\mathbb{P}^2$ passing through a fixed point.

The Roman surface can be inscribed in a regular tetrahedron; the face of the tetrahedron opposite the distinguished pole is tangent to the surface along a circle whose axis is an axis of threefold symmetry.

Choose a new orthonormal frame of reference $(0, X, Y, Z)$ in which the pole is at the origin, the tangent plane at the pole is

$$Z = 0,$$

and the opposite face of the circumscribed tetrahedron is

$$Z = 1.$$

The circle of contact of the Roman surface with the plane $Z = 1$ now has center $(0, 0, 1)$ and radius $\sqrt{2}/3$.

Each ellipse passing through the pole thus touches the two planes $Z = 0$ and $Z = 1$ at two points which define a diameter of the ellipse: one of these points traces out the circle with center $(0, 0, 1)$ and radius $\sqrt{2}/3$, and the other is the point 0. The envelope of the planes of these ellipses is a cone which intersects the plane $Z = 1$ in a threecuspidal hypocycloid circumscribed about the above circle.

We have singled out one particular diameter for each of the ellipses passing through the pole; the length of this diameter is constant and is equal to $\sqrt{11}/3$ while the conjugate diameter has length $2/3$. Let us denote by $\vec{j}$ and $\vec{k}$ vectors representing these two diameters, such that $\vec{k}$ is parallel to the tangent to the hypocycloid and $\vec{j}$ is the position vector of a point of the circle in the plane $Z = 1$ with respect to the origin 0 (Figs. 31, 32).

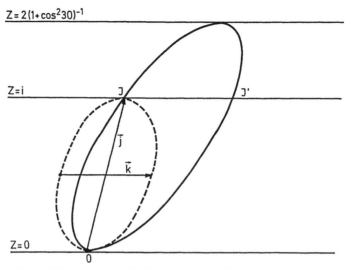

Fig. 32 Dashed line: ellipse generating the Roman surface. Continuous line: ellipse generating the Boy surface.

It is a property of the hypocycloid that the polar angle of the tangent is half of the polar angle (measured in the opposite sense) of the endpoint of the corresponding $\vec{j}$ lying on the circle inscribed in the hypocycloid (Fig. 31). So we can write

$$\vec{j} = \begin{bmatrix} \sqrt{2/3} \cdot \cos 2\theta \\ \sqrt{2/3} \cdot \sin 2\theta \\ 1 \end{bmatrix} \qquad \vec{k} = \begin{bmatrix} 2/3 \cdot \cos \theta \\ -2/3 \cdot \sin \theta \\ 0 \end{bmatrix}$$

We can then parametrize each ellipse θ = constant by associating to the point M the parameter t of the point of intersection of the line OM with the tangent to the hypocycloid parametrized by (Fig. 32)

$$t \mapsto \vec{j} + t\vec{k}$$

This tangent is parallel to $\vec{k}(\theta)$. We obtain the following parametrization of the Roman surface

$$\begin{bmatrix} X \\ Y \\ Z \end{bmatrix} = (1+t^2)^{-1} \cdot (\vec{j}(\theta) + t\vec{k}(\theta)) \qquad 0 \leq \theta < 2\pi \quad t \geq 0$$

It highlights in a striking way the properties we have discussed; we can interpret θ as giving the longitude on the sphere, which labels the meridians which are mapped onto the ellipses of the Roman surface passing through the distinguished poles; the parameter t is the tangent of the latitude.

2.4 A Boy Surface of Degree Six

To deform the Roman surface in a way which respects the geometric requirements of being compatible to the Boy surface, it will be sufficient to replace the term $(1+t^2)^{-1}$ by a suitable multiplicative factor $f(\theta, t)$. We want the locus of points M given by

$$\overrightarrow{OM} = f(\theta, t) \cdot (\vec{j}(\theta) + t\vec{k}(\theta)),$$

where θ is constant, to be a conic which is tangent at O to the plane $Z = 0$. The function f must, therefore, be of the form

$$f(\theta, t) = (a(\theta) - tb(\theta) + t^2 c(\theta))^{-1}$$

where for all θ the discriminant $b^2 - 4ac$ is strictly less than zero. For $t = 0$, we have $\overrightarrow{OM} = \vec{j}(\theta)$ which implies that

$$a(\theta) = 1.$$

We know that for the Boy surface the height function Z has a minimum, three saddles and three maxima (sec. 2.2). These critical points correspond to places at which both the derivatives $\partial_\theta M_3$ and $\partial_t M_3$ vanish simultaneously; here M_3 denotes the Z coordinate of M. We observe that the critical points which are not extremal, i.e. the saddles, occur when $t = 0$, in other words

$$M_3 = 1$$

Boy's drawings [BOY] show that the plane of the saddles meet the surface in a closed curve with three double points. This suggests that we examine the section by the plane $Z = 1$. The ellipse with parameter θ intersects the plane $Z = 1$ at the points for which

$$f(\theta, t) = 1,$$

that is to say

$$t(b - ct) = 0.$$

The solution $t = 0$ gives the point J such that $\overrightarrow{OJ} = \vec{j}(\theta)$, and $t = b(\theta) c(\theta)^{-1}$ gives the point J' with

$$\overrightarrow{OJ'} = \vec{j}(\theta) + b(\theta) c(\theta)^{-1} \cdot \vec{k}(\theta)$$

To follow Boy's indications, we will choose b and c in such a way that the point J' describes an *elongated hypocycloid*. This curve is the locus of a fixed point of the extensions of a radius R of a circle which rolls without slipping around the inside of another fixed circle of radius $3R$ – the fixed point lies outside the rolling circle (Fig. 33).

The quantity $2/3 \cdot b(\theta) c(\theta)^{-1}$ represents the signed length which must be measured out along the tangent to the hypocycloid starting in the direction of $\vec{k}(\theta)$ from the point J. If this quantity vanishes, J' coincides with J, and so J' describes the circle inscribed in the hypocycloid; this gives the Roman surface. If the length is $2\sqrt{2}/3 \cdot \cos 3\theta$, the point J' traces out the hypocycloid. For J' to describe the elongated hypocycloid as desired, we choose a quantity that has a phase difference of $\pi/2$ with $\cos 3\theta$, which means

$$b(\theta) c(\theta)^{-1} = d \sin 3\theta$$

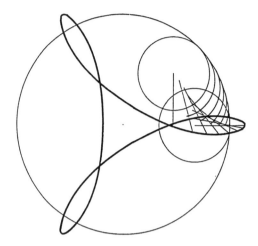

Fig. 33 A mechanism for tracing the elongated hypocycloid.

Bearing in mind the condition

$$b^2(\theta) - 4c(\theta) < 0$$

we choose

$$b(\theta) = \sqrt{2} \sin 3\theta \qquad c(\theta) = 1.$$

We obtain the following parametrization

$$\overrightarrow{OM} = (1 - \sqrt{2}\, t \sin 3\theta + t^2)^{-1} \cdot (\vec{j}(\theta) + t\vec{k}(\theta))$$

where $\theta \in [0, \pi]$ and $t \in [-\infty, +\infty]$ (**Plates 39–43**). Setting $t = \tan \varphi$ we have:

Proposition 1.

If θ and φ are longitude and latitude on the sphere measured with respect to the two poles, then factoring through the antipodal relation the parametrization

$$M(\theta, \varphi) = \cos\varphi \cdot (1 - \sqrt{2} \sin\varphi \cos\varphi \sin 3\theta)^{-1} \cdot \begin{bmatrix} (\sqrt{2}/3) \cos\varphi \cos 2\theta + (2/3) \sin\varphi \cos\theta \\ (\sqrt{2}/3) \cos\varphi \sin 2\theta - (2/3) \sin\varphi \sin\theta \\ \cos\varphi \end{bmatrix}$$

defines a C^1-immersion of $\mathbb{P}^2$ in $\mathbb{R}^3$. This immersion is smooth everywhere except at the point corresponding to the two poles at which it is not of class C^2. The curve of self-intersection can be parametrized by

$$N(\theta) = 2(3 - \cos 6\theta)^{-1} \cdot \begin{bmatrix} (2/3) \sin(\pi/4 - 3\theta) \cos\theta \\ (2/3) \sin(\pi/4 - 3\theta) \sin\theta \\ 1 \end{bmatrix}$$

2.4 A Boy Surface of Degree Six

This is a sextic curve which is of type direct three-bladed propeller. All the other conditions for this to be a direct Boy surface in the sense of section 2.2 are also satisfied.

The details of the proof are given in [AP]. It is now possible to define a *direct Boy surface* to be the image of a C^1-immersion of $\mathbb{P}^2$ in $\mathbb{R}^3$ which is C^1-ambient isotopic to the image of the immersion we have just defined (**Plates 40, 41, 44–47**).

We remark that the level curves of constant Z conform to the description given by Boy (**Plates 48–50**).

Furthermore, the Boy surface we have constructed here has the remarkable property of being the whole of the real zero set of a sextic polynomial. The Boy surface therefore appears as a *rational* real algebraic surface (i.e. one which can be parametrized by rational functions); it has degree six and is genereated by a one parameter family of conics passing through a fixed point:

Proposition 2.

The direct Boy surface defined in proposition 1 is the real zero set of the polynomial

$$p(X, Y, Z) = 64(1-Z)^3 Z^3 - 48(1-Z)^2 Z^2 (3X^2 + 3Y^2 + 2Z^2) + 12(1-Z) Z (27(X^2 + Y^2)^2$$
$$- 24 Z^2 (X^2 + Y^2) + 36\sqrt{2}\, YZ (Y^2 - 3X^2) + 4Z^4) + (9X^2 + 9Y^2 - 2Z^2)(-81(X^2 + Y^2)^2$$
$$- 72 Z^2 (X^2 + Y^2) + 108\sqrt{2}\, XZ (X^2 - 3Y^2) + 4Z^4).$$

Proof.

We first note that the polynomial $p(X, Y, Z)$ does indeed vanish on the image of the immersion defined in proposition 1.

Let us now investigate the possible existence of a zero of p which lies outside the image of the parametrization (such points do exist, for example, for the surface of sec. 2.2 proposition 4). We replace θ and φ by new variables

$$u = \cos\theta \cos\varphi \quad v = \sin\theta \cos\varphi \quad w = \sin\varphi \cos\varphi$$

so as to obtain the rational parametrization

$$F(u, v, w) = ((u^2 + v^2 + w^2)(u^2 + v^2) - \sqrt{2}\, vw(3u^2 - v^2))^{-1} \cdot \begin{bmatrix} \sqrt{2}(u^2 + v^2)(u^2 - v^2 + \sqrt{2}uw)/3 \\ \sqrt{2}(u^2 + v^2)(2uv - \sqrt{2}vw)/3 \\ (u^2 + v^2)^2 \end{bmatrix}$$

This can be interpreted as a rational mapping from $\mathbb{P}^2$ to $\mathbb{R}^3$ if we factor through the quotient defined by the equivalence relation (sec. 1.2 rep. 4)

$$(u, v, w) \sim (u', v', w') \quad \text{if} \quad u/u' = v/v' = w/w'.$$

A point of $\mathbb{P}^2$ is then specified by homogeneous coordinates (u, v, w).

It is clear that if a real zero of p is not in the image of F, then it must be the image of two complex conjugate points of $\mathbb{P}^2(\mathbb{C})$. It is therefore a singular point M of the complex algebraic surface

$$p(X, Y, Z) = 0$$

The planes of the ellipses generating the Boy surface fill out the whole of $\mathbb{R}^3$; in particular there is one of these planes passing through the point M. This plane intersects the surface in a conic of the generating family and a quartic curve with M as a singular point.

The ellipse is the image under F of the projective line

$$u \sin \theta - v \cos \theta = 0$$

where θ is the longitude of the meridian of the sphere corresponding to the line. To determine the remainder of the intersection of the surface with the plane of the ellipse, we find the intersection of the plane with the other ellipses. This is the unicursal quartic which is the image of the projective line

$$w - \sqrt{2}\, u \cos 2\theta + \sqrt{2}\, v \sin 2\theta = 0.$$

The only real singular points of this quartic are an isolated *tacnode* (a double point at which the two tangents coincide) and an ordinary double point located on the curve of self-intersection of the Boy surface. We have shown that the point M lies in the image of the parametrization F, contradicting our hypothesis about M.

Thus every real zero of p is in the image of F. q.e.d.

We observe that if we regard the self-intersection set of the Boy surface as a complex algebraic curve of degree six, we find a triple point at $(0, 0, 2/3)$, which is the triple point of the surface, and an isolated real double point at the pole $(0, 0, 0)$. The tangents at $(0, 0, 0)$ are the two complex conjugate lines

$$X + iY = Z - 1 = 0 \qquad X - iY = Z - 1 = 0.$$

The singular set of the complex algebraic surface defined by p is the union of the sextic curve described above and its two tangents at the pole.

There is an obvious way to perform the deformation of the Roman surface into the Boy surface; this is to introduce a deformation parameter d into the factor $f(\theta, t)$ by writing

$$f_d(\theta, t) = (1 - td\sqrt{2} \sin 3\theta + t^2)^{-1}$$

We can then show [AP]:

Proposition 3.

We can transform the Roman surface (given by $d = 0$) into the Boy surface (for which $d = 1$) by the one parameter family of mappings of $\mathbb{P}^2$ in $\mathbb{R}^3$ defined by

$$F_d(\theta, t) = (1 - td\sqrt{2} \sin 3\theta + t^2)^{-1} \cdot \begin{bmatrix} (\sqrt{2}/3) \cos 2\theta + (2/3) t \cos \theta \\ (\sqrt{2}/3) \sin 2\theta - (2/3) t \sin \theta \\ 1 \end{bmatrix}$$

The six Whitney umbrellas of the Roman surface disappear simultaneously; hyperbolic confluences of three pairs occur at the points

$$(\theta, t) = (\pi/9, 1/\sqrt{2}), \quad (7\pi/9, 1/\sqrt{2}), \quad (13\pi/9, 1/\sqrt{2})$$

when the deformation parameter takes the value $d = 1/\sqrt{3}$.

2.4 A Boy Surface of Degree Six

The main stages of this deformation are illustrated in **Plates 40, 41, 51–57**. Every individual surface with d constant is a rational algebraic surface of degree six, except when $d = 0$; for this value of d the surface decomposes into the Roman surface and its tangent plane (counted twice) at the distinguished pole.

Each of these surfaces is generated by ellipses passing through the pole $(0, 0, 0)$ which is fixed throughout the deformation. The triple point is also fixed, as are the tangent plane at the pole, the planes of the ellipses and the axis of threefold symmetry.

Let us point out some of the features of the deformation of the set of singular points of these surfaces. Starting with $d = 1$, we look at what happens to the algebraic curve of degree six containing the self-intersection set of the Boy surface. The blades of the propeller grow as d decreases; at $d = 1/\sqrt{3}$ **(Plate 53)** they cease to be in the image of F_d and they then become tangent to the plane at infinity. Contrary to what one might expect from a superficial examination, this occurs before d reaches the value zero: in fact it happens at $d = (\sqrt{2} - 1)^2$. As d continues to decrease, the blades pass through the plane at infinity to form three new arcs **(Plate 55)**. When d finally reaches zero, the new arcs join up with the other branches of the curve passing through the triple point. The configuration for $d = 0$ consists of the three double lines of the Roman surface plus a nonsingular cubic curve in the double tangent plane at the pole **(Plates 56, 57)**.

Chapter 3

More about Immersions in the 3-dimensional Sphere

3.1 Self-Transversal Immersions of the Real Projective Plane

There are indeed other ways of immersing the projective plane in $\mathbb{R}^3$ apart from that of Boy; we shall investigate them in the sections below.

Theorem 5 of section 2.1 gave a classification of smooth immersed closed surfaces in $\mathbb{R}^3$ which left essentially only two possibilities for $\mathbb{P}^2$, namely the Boy surface and its mirror image. These can be defined explicitly by taking the parametrizations F_1 and F_{-1} of section 2.4 and by making a slight modification so as to render them smooth at the pole.

However, we are going to introduce a classification of immersions which is finer than the one in section 2.1 given by regular homotopy. It is this new classification which will enable us to distinguish new immersions of $\mathbb{P}^2$ in $\mathbb{R}^3$.

We will use the idea of a transverse immersion, and so we will first need to define what we mean by the *tangent bundle* of a smooth manifold M without boundary.

Consider the case of an open subset U of $\mathbb{R}^n$. The vectors tangent to U at a particular point form a vector space which is naturally isomorphic to $\mathbb{R}^n$, which is denoted by $T_a U$ and is called the *tangent space* to U at a. If, however, we want to be able to distinguish between tangent vectors at two different points a and b, we have to define the idea of a *tied vector:* this is a pair (a, v) where a is a point and v a vector. If v is a vector in the tangent space $T_a U$, we say that (a, v) is a tangent vector to U at a.

The *tangent bundle* to U is then the set of all tangent vectors to U, in other words the disjoint union

$$\bigcup_{a \in U} T_a U$$

equipped with the structure of a smooth $2n$-dimensional manifold given by the product $U \times \mathbb{R}^n$ (prop. 2 sec. 2.1).

When M is a smooth manifold without boundary of dimension n, we must take extra precautions to insure that the definition of a tangent vector does not depend on the choice of local coordinate system.

Suppose u and v are two vectors in $\mathbb{R}^n$, and $U \xrightarrow{h} hU$ and $V \xrightarrow{k} kV$ are two systems of local coordinates for M at a. Then we say that (h, u) is equivalent to (k, v) if

$$v = D(kh^{-1})(ha) \cdot u .$$

3.1 Self-Transversal Immersions of $\mathbb{P}^2$

An equivalence class is written $[a, h, u]$ and is called a *tangent vector to M at a*. The set of tangent vectors to M at a has a natural vector space structure of dimension n given by:
$$\lambda [a, h, u] = [a, h, \lambda u]$$
$$[a, h, u] + [a, k, v] = [a, k, v + D(kh^{-1})(ha) \cdot u].$$

This space is called the *tangent space to M at a* denoted by $T_a M$. We do indeed recover the previous definition of the tangent space $T_a U$ where U is an open subset of $\mathbb{R}^n$, since the existence of a canonical system of local coordinates on U gives an explicit isomorphism between $T_a U$ and $\mathbb{R}^n$.

We define an atlas on the disjoint union
$$TM = \bigcup_{a \in M} T_a M$$
by starting from an atlas on M given by homeomorphisms
$$U_i \xrightarrow{h_i} h_i U_i.$$
Here the U_i are open sets whose union is M and their images $h_i U_i$ are open subsets of $\mathbb{R}^n$. Extend the maps h_i so as to get mappings
$$\bigcup_{a \in U_i} T_a M \xrightarrow{Th_i} T(h_i U_i)$$
such that
$$Th_i [a, h, u] = [h_i a, h h_i^{-1}, u].$$

The tangent bundles $T(h_i U_i)$ are clearly open subsets of $\mathbb{R}^{2m}$ and the maps Th_i define an atlas for a smooth structure on TM. This structure does not depend on the choice of atlas for M, and makes TM into a smooth $2n$-dimensional manifold without boundary which is called the *tangent bundle* to M [SP].

In general, the tangent bundle TM is non-trivial, that is to say it is not given by the product $M \times \mathbb{R}^n$; although this can happen in some cases, for example, when M is the circle S^1 or an open subset of $\mathbb{R}^n$.

Example 1.

In each of the tangent planes to the unit sphere in $\mathbb{R}^3$ we can define a Euclidean structure: we fix the origin to be at the point of contact a, and we define a scalar product $\langle u, v \rangle$ using the scalar product operation in the ambient space. The disjoint union of these tangent planes is the tangent bundle TS^2 to the sphere.

The mapping
$$(a, u, v) \mapsto \langle u, v \rangle$$
from $S^2 \times T_a S^2 \times T_a S^2$ to $\mathbb{R}$ is smooth: we have constructed what is called a *Riemannian metric* on S^2. Using this metric we can define the *unit tangent bundle* of S^2 to be the disjoint union of the unit circles in the tangent planes $T_a S^2$. We give this bundle the smooth structure which makes the canonical injection into TS^2 into a smooth embedding.

The same construction on the circle S^1 would give a unit tangent bundle which is the product $S^1 \times S^0$; in other words two circles.

To understand the topology of the unit tangent bundle of S^2, we consider the map H which sends a point (x, y, z, t) of S^3 to

$$H(x, y, z, t) = (x^2 + y^2 - z^2 - t^2,\ 2xz + 2yt,\ 2xt - 2yz).$$

Since (x, y, z, t) is in S^3 (this means $x^2 + y^2 + z^2 + t^2 = 1$), its image is a point of S^2. The mapping H is a surjection and is called the *Hopf fibration*. The inverse images $H^{-1}(a, b, c)$ of the points of S^2, which we call the fibres, are circles S^1. For the fibre passing through (x, y, z, t) can be parametrized by

$$(x \cos u - y \sin u,\ x \sin u + y \cos u,\ z \cos u - t \sin u,\ z \sin u + t \cos u)$$

as u varies through $S^1 = \mathbb{R}/2\pi\mathbb{Z}$.

We now see that the inverse image under H of the domain of definition U of a system of local coordinates on S^2 can be identified with $U \times S^1$ which is the unit tangent bundle of U. We can thus prove that the unit tangent bundle of S^2 can be identified with S^3. Therefore the unit tangent bundle TS^2 is non-trivial; otherwise it would be the product $S^2 \times S^1$; but $S^2 \times S^1$ is not homeomorphic to S^3 because the fundamental group of S^3 is trivial while that of $S^2 \times S^1$ is (sec. 1.2 ex. 3 and 4)

$$\pi(S^2 \times S^1) = \pi(S^2) \times \pi(S^1) = 1 \times \mathbb{Z} = \mathbb{Z}.$$

As we now have available the concept of the tangent bundle, we can easily introduce that of a derivative. If $M \xrightarrow{f} N$ is a C^r-mapping of two smooth manifolds without boundary, and if $U \xrightarrow{h} hU$ and $V \xrightarrow{k} kV$ are local coordinate systems at a and $f(a)$, then we put

$$f_*[a, h, u] = [f(a), k, D(kfh^{-1})(ha) \cdot u].$$

This gives the definition of the derivative as a mapping $TM \xrightarrow{f_*} TN$ of class C^{r-1} which is independent of the choice of local coordinate systems. It also satisfies the following equalities:

$$(1_{M^*}) = 1_{TM} \quad (g \circ f)_* = g_* \circ f_* \quad (f_{|U})_* = (f_*)_{|TU}$$

where $f_{|U}$ denotes the restriction of f to an open subset U of M. We also remark that f_* restricts to a linear map from $T_a M$ to $T_{f(a)} N$; if we choose local coordinate systems at a and $f(a)$, these determine basis for $T_a M$ and $T_{f(a)} N$ in which this linear map is given by the jacobian matrix of f.

When we identify the tangent bundle to an open subset U of $\mathbb{R}^m$ with the product $U \times \mathbb{R}^m$, we recover the usual definition of the derivative of a map $U \xrightarrow{f} \mathbb{R}^n$.

Let us now consider two smooth manifolds without boundary M and N and a C^1-immersion $M \xrightarrow{f} N$; and let us suppose that M is compact. Since an immersion is locally an embedding, the preimages of a mutiple point of f are isolated, and since M is compact, they are finite in number. The derived mapping f_* sends the tangent space $T_a M$ bijectively and linearly onto a subspace of the tangent space $T_{f(a)} N$.

We say that a C^1-immersion $M \xrightarrow{f} N$ is *self-transversal* if, at each multiple point b the images of the tangent spaces at the preimages of b intersect in a subspace of $T_b N$ of

3.1 Self-Transversal Immersions of $\mathbb{P}^2$

least possible dimension; if the preimages of b are $a_1, \ldots, a_p$, and if $\dim M = m$ and $\dim N = n$, we require that

$$\dim f_* T_{a_1} M \cap \ldots \cap f_* T_{a_p} M = pm + (1-p) n .$$

Example 2.

The following parametrization of the lemniscate of Bernoulli

$$x = \sin 2t \cdot (1 + \cos^2 t)^{-1}$$
$$y = 2 \sin t \cdot (1 + \cos^2 t)^{-1}$$

where $t \in S^1 = \mathbb{R}/2\pi \mathbb{Z}$ defines a self-transversal immersion of S^1 in $\mathbb{R}^2$. The two tangents at the double point intersect transversally (Fig. 34).

Fig. 34 Lemniscate of Bernoulli.

The *paquerette de Mélibée* can be parametrized by

$$x = \sin 3t \cdot \cos t$$
$$y = \sin 3t \cdot \sin t$$

where $t \in S^1 = \mathbb{R}/\pi \mathbb{Z}$; this is a smooth immersion of S^1 in $\mathbb{R}^2$ which is not self-transversal because of the existence of a triple point at the origin (Fig. 35).

Fig. 35 Pâquerette de Mélibée.

Example 3.

We consider the parametrization of the Klein bottle defined in section 2.1 example 14. Recall that this is given by

$$x = \cos 2t \cdot \sin u / (1-w)$$
$$y = (\sin 2t \cdot \sin u - \sin t \cdot \cos u) / (\sqrt{2}(1-w))$$
$$z = \cos t \cdot \cos u / (1-w)$$

where

$$w = (\sin t \cdot \cos u + \sin 2t \cdot \sin u) / \sqrt{2}$$

We have seen that these formulae give a smooth immersion of the Klein bottle $\mathbb{K}$ in $\mathbb{R}^3$ whose image is the real algebraic surface with equation

$$(x^2 + y^2 + z^2 + 2y - 1)((x^2 + y^2 + z^2 - 2y - 1)^2 - 8z^2) + 16xz(x^2 + y^2 + z^2 - 2y - 1) = 0.$$

The multiple points of this immersion are all double points; they form a circle given by

$$z = 0 \qquad x^2 + (y-1)^2 = 2.$$

To investigate whether the immersion is self-transversal, we take a new frame of reference in which the origin lies on the circle:

$$x = \sqrt{2} \cos v + X \cos v - Y \sin v$$
$$y = 1 + \sqrt{2} \sin v + X \sin v + Y \cos v$$
$$z = Z.$$

We calculate the restriction to the XOZ plane of 2-jet at the origin of the equation of the algebraic surface:

$$32((X^2 - Z^2)(1 + \sqrt{2} \sin v) + 2XZ \cos v).$$

The discriminant of this quadratic form is greater than zero for all values of v; this proves that, for any point O of the circle of self-intersection, the algebraic surface meets the plane XOZ along a curve which has two distinct branches at O.

This shows that the above smooth immersion of $\mathbb{K}$ in $\mathbb{R}^3$ is self-transversal (**Plate 17**).

Example 4.

It is proved in [AP] that the C^1-immersion of Boy given by the map $F_1(\theta, t)$ of sec. 2.4 is self-transversal. The multiple points form a unicursal sextic curve (sec. 2.4 prop. 1); every point of the curve is a double point of the immersion, except for one point which is a triple point both for the curve and for the immersion.

In the two previous examples, the self-transversally immersed surfaces had self-intersection sets consisting of immersed curves. This is a general property; we shall give a proof which works in the smooth case.

If N is a smooth manifold without boundary, then a smooth *submanifold* is defined to be the image of a smooth manifold M under a smooth embedding $M \to N$. If M is a smooth submanifold of N then there is a unique smooth structure on M which makes the canonical injection $M \to N$ into a smooth embedding. In particular, an embedded surface as defined in sec. 2.1 is a smooth submanifold of $\mathbb{R}^3$.

3.1 Self-Transversal Immersions of $\mathbb{P}^2$

Let $M \xrightarrow{f} N$ be a smooth immersion, where the manifolds without boundary M and N have dimensions m and n and let M be compact. We will denote N_k the set of points of N which have at least k preimages under f, and S_k will denote the subset of $M^k = M \times \ldots \times M$ consisting of k-tuples of distinct points of M such that all k points have the same image under f. Projection onto the first factor will be denoted by $M^k \xrightarrow{p} M$.

Proposition 1.

Let $M \xrightarrow{f} N$ be a smooth self-transversal immersion. Then if S_k is nonempty it is a smooth submanifold of M^k without boundary; it is compact and has dimension $km + (1-k)n$. The restriction to S_2 of the projection p is a self-transversal immersion of S_2 in M, whose image under f is $fp S_2 = N_2$.

Proof.

Let $(x_1^i, \ldots, x_k^i)$ be a sequence in S_k which converges to a point $(x_1, \ldots, x_k)$ in the compact set M^k. Then by continuity of f

$$f(x_1) = \ldots = f(x_k)$$

and since f is locally injective (as is any immersion), we know that $(x_1, \ldots, x_k)$ lies in S_k. We have shown that S_k is a closed subset of M^k, and therefore it is compact.

Consider the mapping $\bar{M} \xrightarrow{fx \ldots xf} N^k$ where $\bar{M} = M \setminus C$ and C is the set of k-tuples having at least two components equal and where

$$fx \ldots xf(x_1, \ldots, x_k) = (f(x_1), \ldots, f(x_k)).$$

The image of this mapping is contained in the diagonal D of N^k; the diagonal consists of k-tuples all of whose components are equal, and it is a smooth submanifold of N^k of codimension $(k-1)n$.

The self-transversality condition on the immersion f means that, at each point $(f(x_1), \ldots, f(x_k))$ of D, the tangent space to N^k is spanned by the tangent space to D and the image under $f_* x \ldots x f_*$ of the tangent space to $\bar{M}$ at $(x_1, \ldots, x_k)$. This shows that the inverse image S_k of D by the map $fx \ldots xf$ is a smooth submanifold of $\bar{M}$, and hence of M^k. The codimension of S_k is equal to that of D (see e.g. [HI]):

$$\dim S = km + (1-k)n.$$

We now consider a point x of M which has l preimages in S_2 under the map p, say

$$(x, x_1), \ldots, (x, x_l).$$

Let $T_1, \ldots, T_l$ denote the tangent spaces to S_2 at these points; then the image of T_i by the derivative $(fp)_* = f_* p_*$ is equal to the intersection of the images under f_* of the tangent spaces $T_x M$ and $T_{x_i} M$ at the two points x and x_i. Since f is an immersion,

$$\dim p_* T_1 \cap \ldots \cap p_* T_l = \dim f_* p_* T_1 \cap \ldots \cap f_* p_* T_l$$
$$= \dim f_* T_x M \cap f_* T_{x_1} M \cap \ldots \cap f_* T_x M \cap f_* T_{x_l} M.$$

As any of the points $x, x_1, \ldots, x_l$ are distinct and as f is a self-transversal immersion, we have

$$\dim p_* T_1 \cap \ldots \cap p_* T_l = \dim f_* T_x M \cap f_* T_{x_1} M \cap \ldots \cap f_* T_{x_l} M$$
$$= (l+1)m - ln = l \dim S_2 + (1-l) \dim M.$$

So p, therefore, restricts to a self-transversal immersion of S_2 in M. It is clear that $fpS_2 = N_2$.
q.e.d.

Example 4 shows that in general the projection p does not restrict to a self-transversal immersion of S_3 in M: in that example S_3 consists of six points which project in pairs onto three points of $\mathbb{P}^2$. In this case the idea of an immersion is not very meaningful, but the self-transversality condition, expressed as a condition on dimensions, would require that p has no double points on the 0-dimensional manifold S_3.

When M is the projective plane $\mathbb{P}^2$ and N is the three dimensional sphere S^3, the above result tells us that a self-transversal immersion $\mathbb{P}^2 \to S^3$ has no quadruple points and at most finitely many triple points. The triple points lie on a smooth curve of double points, and this curve is itself the image of another self-transversal immersion.

By a particular case of Sard's theorem [HI], the complement of the image of a smooth immersion of $\mathbb{P}^2$ in S^3 is a dense subset of S^3; so by removing a point which is not in the image of $\mathbb{P}^2$, we can assume that we have an immersion of $\mathbb{P}^2$ in $\mathbb{R}^3$. Proposition 7 of sec. 1.1 shows that any such immersion must have a double point.

In fact, a theorem of T. Banchoff [BA] states that any self-transversal immersion of $\mathbb{P}^2$ in S^3 has an odd number of triple points — in particular there is at least one.

Thus, taking the simplest possibility for the self-intersection set of a smooth self-transversal immersion of $\mathbb{P}^2$ in S^3, we find that the set S_2 of preimages of the double points is a closed curve (a circle S^1). Projecting this curve to $\mathbb{P}^2$ gives a self-transversal immersion with three double points which are the preimages of the triple point. This is precisely the case with the Boy immersion, which can be represented by the parametrization F_1 of sec. 2.4 modulo a small perturbation at the pole.

In the next section, we want to classify, according to their general shape, the images of smooth self-transversal immersions of $\mathbb{P}^2$ in S^3 whose self-intersection set is similar to that of the Boy surface.

For this, we need to define a new equivalence relation between smooth mappings of manifolds. We shall be looking at the space of smooth maps from a compact smooth manifold M without boundary to a smooth manifold N without boundary. We will give this set the topology of compact convergence of all partial derivatives for all local coordinate systems on M and N.

An *oriented smooth structure* on M is defined to be any smooth structure induced by an atlas in which all changes of coordinates are given by smooth diffeomorphisms with positive jacobian determinants; if such a structure exists, M is said to be *orientable*.

If for example a smooth manifold has a global system of coordinates, then it is orientable; this is the case with $\mathbb{R}^n$. Also orientability is preserved by taking products, and above all any compact smooth submanifold of $\mathbb{R}^{n+1}$ of dimension n (and without boundary) is orientable [HI]. Thus, example 9 of sec. 2.1 shows that a closed surface which is orientable in the sense of section 1.1 remains orientable according to the above definition.

3.1 Self-Transversal Immersions of $\mathbb{P}^2$

We will say that a diffeomorphism f from a smooth orientable manifold M without boundary to itself is *orientation-preserving* if there is an atlas in which all coordinate change transformations have positive jacobian determinants, and in which f also has positive jacobian determinant.

Example 5.

The antipodal mapping on the unit sphere S^2 in $\mathbb{R}^3$ is not orientation-preserving. For, if we take the atlas consisting of two coordinate systems at $(0, 0, 1)$ and $(0, 0, -1)$, given by

$$h(x, y, z) = (1+z)^{-1} \cdot (x, y)$$
$$k(x, y, z) = (1-z)^{-1} \cdot (-x, y),$$

then the change of coordinates is given by

$$kh^{-1}(u, v) = (u^2 + v^2)^{-1} (-u, v);$$

this does indeed have positive jacobian determinant. Now if f is the antipodal map

$$f(x, y, z) = (-x, -y, -z)$$

then

$$kfh^{-1}(u, v) = (u, -v)$$

which shows that f reverses the orientation of S^2.

If there was an atlas for the projective plane $\mathbb{P}^2$ in which all the coordinate changes had positive jacobian determinants, then it could be lifted to an atlas of S^2 having the same property and which was also invariant under the antipodal map. This would contradict the fact that f is orientation-reversing on S^2. For this reason $\mathbb{P}^2$ is nonorientable according to the above definition, just as it is nonorientable in the sense of section 1.1.

When M is nonorientable, we shall say by abuse of notation that any diffeomorphism of M preserves the orientation of M.

We now define an equivalence relation on the space $S(M, N)$ of smooth maps from M to N. Two mappings f and g in $S(M, N)$ will be called *equivalent* if there are orientation-preserving smooth diffeomorphisms $M \xrightarrow{h} M$ and $N \xrightarrow{k} N$ such that $g = kfh^{-1}$.

Self-transversal immersions have the property of stability for this kind of equivalence; the proof of this result can be found in [GO].

Proposition 2.

Let f be any self-transversal smooth immersion of a compact manifold without boundary into a manifold N without boundary. Then there is some neighborhood of f in $S(M, N)$ consisting of mappings which are all equivalent to f.

We now return to the case of smooth self-transversal immersions of $\mathbb{P}^2$ in S^3. The spirit of the classification we will be studying is that we wish to recognize the simplest possible shapes which can be taken by $\mathbb{P}^2$ in S^3, independently of parametrizations; this explains the diffeomorphism in the source in the above equivalence relation.

90   3 More About Immersions in S^3

In the target we want to get from one shape to another equivalent shape by a smooth deformation of the ambient space, that is by a smooth ambiant isotopy in the sense of sec. 2.2. A result of J. Cerf states that any smooth orientation-preserving diffeomorphism of S^3 is isotopic to the identity [CE]; this motivates our definition of equivalence for studying immersions of $\mathbb{P}^2$ in S^3.

3.2 Classification of Immersed Projective Planes of Boy Type

We define an *immersed projective plane of Boy type* to be an equivalence class $[f]$ in the sense of sec. 3.1, where f is a smooth self-transversal immersion of $\mathbb{P}^2$ in S^3 whose self-intersection set satisfies the following condition:

(*) the image under fp of the curve S_2 of pairs of inverse images of the double points of f is a *direct three-bladed propeller*, in other words it is smoothly ambient isotopic to the image of the parametrization $N(\theta)$ of prop. 1 sec. 2.4 (**Plates 42, 43**). Here the mapping p is projection onto the first factor from $\mathbb{P}^2 \times \mathbb{P}^2$ to $\mathbb{P}^2$.

This condition means that $[f]$ has precisely one triple point, and since S_2 is a double cover of the direct three-bladed propeller consisting of double points, S_2 has at most two connected components.

If we look at immersions of $\mathbb{P}^2$ in $\mathbb{R}^3$ rather than S^3, this amounts to fixing a point ∞ of S^3, which is not in the image of the immersion, and projecting stereographically with the point ∞ as pole to get $\mathbb{R}^3$. If we consider the Boy immersion given by the parametrization F_1 (sec. 2.4), we could imagine a smooth ambient isotopy of S^3 which stretches the disk bounded by one of the blades of the propeller until it passes through the fixed point ∞ (Fig. 36). This deformation has no effect on the classification for S^3,

Fig. 36 Surface obtained by stretching the disk bounded by one of the blades of the self-intersection curve of the Boy surface, until it passes through the point at infinity.

3.2 Immersed Projective Planes of Boy Type

but introduces extra complications to the classification for $\mathbb{R}^3$. This is the reason why we have chosen S^3 as target instead of $\mathbb{R}^3$.

We now intend to prove that there are precisely two immersed projective planes of Boy type. First we need to give a sketch of some of the properties of links.

We say that two smooth immersions $[a, b] \xrightarrow{\gamma} \mathbb{R}^n$ and $[c, d] \xrightarrow{\delta} \mathbb{R}^n$ are equivalent if there is a smooth order-preserving diffeomorphism $[a, b] \xrightarrow{h} [c, d]$ such that $\delta h = \gamma$. An equivalence class is called a *smooth path* in $\mathbb{R}^n$. The starting point, the endpoint, the inverse smooth path, simple smooth paths, and smooth loops are defined as in sec. 1.1.

Similarly, recalling the definition of the sum of two paths in sec. 1.1, any sum of finitely many smooth paths is called a *piecewise smooth path;* its image in $\mathbb{R}^n$ is called the *support* of the piecewise smooth path. At any point of such a path which has only one preimage we can define two *half-tangents* – except of course at the two extremities where there is only one.

Consider two piecewise smooth paths in $\mathbb{R}^2$; we shall say that they meet transversally if the following conditions hold (Fig. 37):

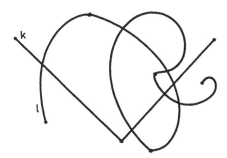

Fig. 37 Two piecewise smooth paths meeting transversally.

(i) there are parametrizations $I \xrightarrow{k} \mathbb{R}^2$ and $I \xrightarrow{l} \mathbb{R}^2$ of the two paths such that, for some subdivision

$$0 = t_0 < \ldots < t_n = 1$$

of I, each of k and l restricts to a smooth immersion on each interval $[t_i, t_{i+1}]$ of the subdivision;

(ii) Each point, lying in both supports, is not the image of a t_i under the parametrizations k and l and has precisely one inverse image under each of k and l; the two corresponding tangents are required to be independent.

A *link* is a pair of piecewise smooth simple loops $I \xrightarrow{\gamma} \mathbb{R}^3$ and $I \xrightarrow{\delta} \mathbb{R}^3$ with disjoint supports for which there is a smooth isotopy $\mathbb{R}^3 \times I \xrightarrow{F} \mathbb{R}^3$ such that the two loops

$$k(\,.\,) = (F_1(\gamma(\,.\,), 1), F_1(\delta(\,.\,), 1)) \quad l(\,.\,) = (F_2(\gamma(\,.\,), 1), F_2(\delta(\,.\,), 1))$$

meet transversally; here F_1 and F_2 are the first two components of F.

If the point m is in the supports of both k and l, then their is a neighborhood V of m in $\mathbb{R}^2$ and a smooth isotopy

$$\mathbb{R}^3 \times I \xrightarrow{G} \mathbb{R}^3$$

with the following properties. We have

$$G(V \times \mathbb{R}, 1) = \bar{B}^2 \times \mathbb{R},$$

and the restrictions of the two loops γ and δ to $V \times \mathbb{R}$ are isotopped either to the pair of oriented line segments

$$[(-1, 0, 1), (1, 0, 1)] \qquad [(0, -1, -1), (0, 1, -1)]$$

or to the pair of oriented line segments

$$[(-1, 0, -1), (1, 0, -1)] \qquad [(0, -1, 1), (0, 1, 1)].$$

In the first case we assign to the point m a weight of $1/2$ and in the second case $-1/2$. The sum of the weights of all the points of intersection of the supports of k and l is independent of the choice of the isotopies F and G and is called the *linking number* of the link (γ, δ) (Fig. 38).

The linking number is an integer and is invariant under a *homotopy of links*, in other words a family of links depending continuously on one parameter.

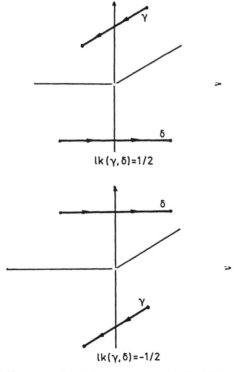

Fig. 38 Two examples of the computation of the linking number.

3.2 Immersed Projective Planes of Boy Type

If two loops are the boundaries of disjoint disks in $\mathbb{R}^3$, then the link is said to be *trivial;* the linking number is equal to zero. The converse however, is false.

Example 1.

The *Whitehead link* defined by Fig. 39 is nontrivial although its linking number is zero.

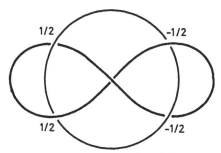

Fig. 39 The Whitehead link.

We note that a link is trivial if and only if it is homotopic as a link to a pair of disjoint coplanar circles (for further details about knots and links see e.g. [GRAM]). We now continue our study of immersions of $\mathbb{P}^2$ in S^3.

Let $[f]$ be an immersed projective plane of Boy type. Denote by

$$S^2 \xrightarrow{q} \mathbb{P}^2$$

the covering of the projective plane by the 2-dimensional sphere (ex. 8 sec. 1.2). The inverse image $C_2 = q^{-1} p S_2$ is the image of a curve without boundary under a smooth self-transversal immersion in S^2.

We define a *tubular neighborhood* V_ϵ of C_2 in S^2 to be the union of the open balls of radius $\epsilon > 0$ with centers on C_2. We take the metric on S^2 induced by that of $\mathbb{R}^3$.

Choose ϵ sufficiently small for C_2 to be a strong deformation retract (see below) of the closure of V_ϵ. A subset A of a topological space X is called a *strong deformation retract* of X if:

(i) there is a continuous map $X \xrightarrow{r} A$, known as *retraction*, such that

$$\forall a \in A \qquad r(a) = a$$

(ii) there is a homotopy $X \times I \xrightarrow{F} X$ such that

$F(x, 0) = x$ $\qquad \forall x \in X$
$F(x, 1) \in A$ $\qquad \forall x \in X$
$F(x, t) = t$ $\qquad \forall x \in A \quad \forall t \in I$

The image of V_ϵ by q will be denoted by W_ϵ. There is now a *smooth vector bundle of dimension 1 with base S_2*, written $E \xrightarrow{\pi} S_2$, and a surjective smooth local diffeomorphism η from E to W_ϵ such that the following diagram commutes

By local diffeomorphism we mean a mapping which restricts, on some neighborhood of each point, to a diffeomorphism.

To say that $E \xrightarrow{\pi} S_2$ is a smooth vector bundle of dimension one with base S_2, means that there is an atlas for S_2 consisting of local coordinate systems $U_i \xrightarrow{h_i} h_i U_i$ such that for each i there is a corresponding local coordinate system on E with source $V_i = \pi^{-1} U_i$ and target $h_i U_i \times \mathbb{R}$ featuring in a commutative diagram

$$\begin{array}{ccc} V_i & \longrightarrow & h_i U_i \times \mathbb{R} \\ \pi \downarrow & & \downarrow pr_1 \\ U_i & \xrightarrow{h_i} & h_i U_i \end{array}$$

Moreover the fibre $\pi^{-1} x$ above each point x of S_2 is given a vector space structure such that for every U_i containing x the mapping $V_i \to h_i U_i \times \mathbb{R}$ induces an isomorphism of the vector space $\pi^{-1} x$ with $h_i x \times \mathbb{R}$.

Example 2.

There are essentially only two smooth vector bundles of dimension one whose base is the circle S^1. There is a trivial bundle $S^1 \times \mathbb{R} \to S^1$ and there is also the bundle

$$I \times \mathbb{R} /\sim \; \longrightarrow I/\sim \; = S^1 \qquad [u, v] \mapsto [u]$$

where the equivalence relation on $I \times \mathbb{R}$ is

$$(0, v) \sim (1, -v)$$

and where that on I identifies 0 and 1. As was shown in sec. 1.1 example 6, the second bundle is an open Möbius strip.

The boundary γ of W_ϵ is a union of piecewise smooth simple loops. Each double point of the smooth self-transversal immersion $S_2 \xrightarrow{P} \mathbb{P}^2$ generates four points at which γ fails to be smooth, and so γ is the sum of twelve smooth paths $d_1, \ldots, d_{12}$ (this is a non-connected sum; see Fig. 40).

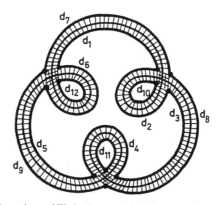

Fig. 40 The boundary of W_ϵ is the sum of twelve smooth paths $d_1, \ldots, d_{12}$.

3.2 Immersed Projective Planes of Boy Type

Without changing the class $[f]$ we may suppose that S^3 minus a point has been identified with $\mathbb{R}^3$ in such a way that the triple point is at the origin. We may also assume that the intersection of the image of $\overline{W}_\varepsilon$ with the cube $C = [-2, 2]^3$ is the union of the strip

$$|x| \leq 1 \qquad |y| \leq 2 \qquad z = 0$$

and the five others obtained by the action of the group $\mathfrak{S}_3$ of permutations of the coordinates (Fig. 41). We can arrange that

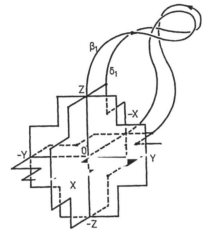

Fig. 41 Neighborhood of the triple point in an immersed projective plane of Boy type.

(i) the intersection of the direct three-bladed propeller fpS_2 with C is the union

$$K = [-X, X] \cup [-Y, Y] \cup [-Z, Z]$$

with

$$X = (2, 0, 0) \qquad Y = (0, 2, 0) \qquad Z = (0, 0, 2).$$

(ii) the images of the twelve points where γ is not smooth are the vertices of the cuboctahedron given by the points for which one coordinate vanishes and the others are ± 1.

The intersection of $f\mathbb{P}^2 \cap C$ will be taken to be the set of points with two coordinates equal to zero and the third lying between -2 and 2.

The remaining part of the three-bladed propeller, i.e. $fpS^2 \setminus C$, consists of three disjoint simple paths $\beta_1, \beta_2, \beta_3$, whose extremities are $X, -X, Y, -Y, Z, -Z$. Suppose that β_1 starts at Z and ends at $-X$. As the three-bladed propeller is direct, the path starting at X (say β_2) finishes at $-Y$; the path β_3 then has starting-point at Y and endpoint at $-Z$ (Fig. 41).

Denote by d_1 the path along the boundary of W_ε whose image δ_1 under the map f begins at $(-1, 0, 1)$ and whose half-tangent at this point is in the direction of the vector $(0, 0, 1)$. This path ends at one of the four points given by (Fig. 41)

$$b_0 = (-1, 0, 1) \quad b_1 = (-1, -1, 0) \quad b_2 = (-1, 0, -1) \quad b_3 = (-1, 1, 0)$$

The endpoint of δ_1 will be denoted by b_i.

If $i \neq 0$,

we construct two simple loops δ_1' and δ_1'' by adding to the two arcs starting at b_i and ending at b_0 which lie along the circle $b_0 b_1 b_2 b_3$.

If $i = 0$,

let $\delta_1' = \delta_1'' = \delta_1$.

Let β_1' be the simple loop which is the sum of the paths

$$\beta_1' = [0, Z] \cdot \beta_1 \cdot [-X, 0] \, .$$

Consider the two linking numbers (β_1', δ_1') and (β_1', δ_1''), and let k be the one with smaller absolute value; and let

$$n_1 = i + 4k$$

where i is the subscript of the endpoint b_i of δ_1 (Fig. 42). Similarly β_2 and β_3 yield integers n_2 and n_3.

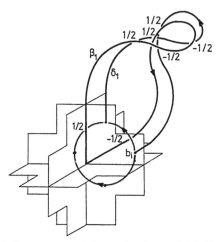

Fig. 42 Computation of the number k: $k = \inf(0, 1) = 0$.

3.2 Immersed Projective Planes of Boy Type

Lemma 1.

All the connected components of E are nonorientable. Furthermore $n_1 + n_2 + n_3 \not\equiv 3 \ (4)$.

Proof.

(i) <u>if S_2 connected</u>

Let δ_2 be the path along the boundary of W_e starting from $b_i + (2, 0, 0)$ whose half-tangent there is the opposite direction to the half-tangent at the end of δ_1. Define δ_3 similarly and put

$$a_0 = b_2 \qquad a_1 = (0, -1, -1) \qquad a_2 = (1, 0, -1) \qquad a_3 = (0, 1, -1).$$

Since S_2 is connected, the subscript j of the endpoint a_j of δ_3 satisfies

$$j \equiv 1 \text{ or } 3 \ (4).$$

If we trace out pS_2 twice, we find

$$2j \equiv 2 \ (4)$$

which proves that E is nonorientable; in fact it is an open Möbius strip as in example 2.

(ii) <u>if S_2 is non connected</u>

As we have seen, S_2 then has two components; the vector bundle E also has two components denoted by E' and E''. Self-transversality of the smooth immersion f means that the images $f_\eta E'$ and $f_\eta E''$ meet transversally along pS_2; so E' and E'' are homeomorphic.

Suppose that E' and E'' define trivial bundles; then by sec. 1.2 prop. 9 the image pS_2 is the union of two loops which are homotopic to ovals in $\mathbb{P}^2$. The self-transversal immersion $S_2 \xrightarrow{P} \mathbb{P}^2$ has three double points while the images of the two connected components of S_2 intersect at an even number of points (cf prop. 2 sec. 2.2). So there are three cases to investigate.

1. The image of one of the components of S_2 has no double points and the image of the other has three double points; the two loops in $\mathbb{P}^2$ are disjoint (Fig. 43).
2. The image of one component of S_2 has no double point and the image of the other has one; these two images meet at two points (Fig. 43).
3. The image of one component of S_2 has two double points and the image of the other has one; the two loops in $\mathbb{P}^2$ are disjoint (Fig. 43).

Looking at a neighborhood of the direct three-bladed propeller, we see the absurdity of these three possibilities. By example 2 the vector bundles E' and E'' are open Möbius strips.
We have

$$n_1 + n_2 + n_3 + 1 \equiv j \ (4).$$

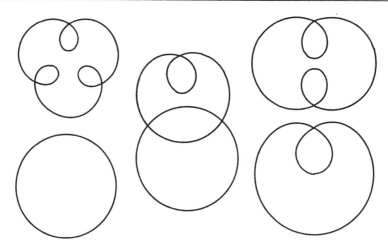

Fig. 43 The three cases for the intersection of the components of pS_2.

In case (i), since j is congruent to either 1 or 3 modulo 4, we get

$$n_1 + n_2 + n_3 \equiv 0 \ (2).$$

In case (ii), since E' and E'' are Möbius strips, j is congruent to 2 modulo 4, which shows that

$$n_1 + n_2 + n_3 \equiv 1 \ (4) \qquad \text{q.e.d.}$$

Lemma 2.

The boundary of W_ϵ is composed of four disjoint simple loops whose images under f form the boundaries of four disjoint disks in $\mathbb{R}^3 \setminus fW_\epsilon$.

Proof.

Each connected component of the boundary of W_ϵ defines the simple loop γ_i which does not meet pS_2. Lemma 1 tells us that pS_2 consists of loops which are homotopic to projective lines in $\mathbb{P}^2$. By prop. 2 sec. 2.2, the loops γ_i are ovals. Corollary 8 sec. 1.2 implies that each loop γ_i bounds a disk in $\mathbb{P}^2$.

The disks bounded by the loops γ_i are disjoint because they are the connected components of the complement of W_ϵ. The hypothesis that pS_2 is a strong deformation retract of W_ϵ ensures that the components of the complement of pS_2 are disjoint open disks; the number of such disks is equal to the number of loops γ_i.

Let us calculate the Euler-Poincaré characteristic according to sec. 1.1; denote by n the number of loops γ_i, which is the number of components of the complement of pS_2.

3.2 Immersed Projective Planes of Boy Type

Since pS_2 has three double points, it decomposes into six edges, and so

$$1 = \chi(\mathbb{P}^2) = n - 6 + 3 .$$

Thus, we do indeed find $n = 4$. It remains only to observe that, on restricting f to each of the four disks bounded by the γ_i, we get a topological embedding which is smooth in the interior of each disk.   q.e.d.

Lemma 3.

Either two of the numbers n_i and n_j are multiples of 4 and the third n_k is not congruent to 3 modulo 4, or

$$n_1 \equiv n_2 \equiv n_3 \equiv 3 \ (4)$$

Proof.

First we remark that the number of components of the boundary of W_e depends only on the residue classes of n_1, n_2, n_3 modulo 4. Also the number of components does not change if the triple (n_1, n_2, n_3) is permuted by an element of $\mathfrak{S}_3$; for a cyclic permutation corresponds to renumbering the blades of the three-bladed propeller, and a transposition amounts to reflecting K in the line

$$x + z = 0 \quad \text{and} \quad y = 0 .$$

Lemma 1 tells us that $n_1 + n_2 + n_3 \not\equiv 3 \ (4)$, and in all remaining cases, the number n of components of the boundary of W_e is given by the table in Fig. 44.   q.e.d.

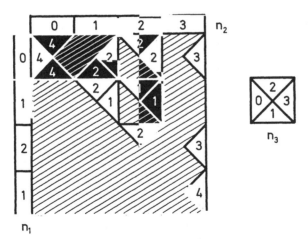

Fig. 44 Table showing the number of boundary components of W_e as a function of the three quantities n_1, n_2, n_3 which take values in $\mathbb{Z}/4$. The square on the right indicates the value of n_3 corresponding to each entry in the table.

Lemma 4.

$|n_1|, \quad |n_2|, \quad |n_3| \leq 3$

Proof.

We showed in lemma 2 that the images under f of the components of the boundary of W_ϵ bound disks which are disjoint from fW_ϵ; So each of these loops, taken together with each of the blades of the three-bladed propeller, forms a trivial link. Because of this constraint, the number k must be equal to either 0 or -1; recall that k was defined above to be whichever of the two linking numbers (β_1', δ_1') and (β_1', δ_1'') has the smaller absolute value.

The same argument shows that if $k = -1$, then $i \neq 0$; so, finally, we have indeed proved that $|n_j| \leq 3$. q.e.d.

Lemma 5.

Up to permutation by an element of $\mathfrak{S}_3$,

$(n_1, n_2, n_3) = (0, 0, 0) \quad \text{or} \quad (0, 0, -2)$

Proof.

We consider the set of loops which are the images under f of the four components of the boundary of W_ϵ, together with the loops formed by the blades of the self-intersection set. We denote by l the sum of the absolute values of the linking number of these loops taken in pairs. Lemma 2 says that if f gives a Boy immersion of the real projective plane, then l must be zero.

The previous lemmae show that, modulo a permutation of the n_i, it is sufficient to discuss the cases in which (n_1, n_2, n_3) is

$(0, 0, -3), \quad (0, 0, -2), \quad (0, 0, 0), \quad (0, 0, 1), \quad (0, 0, 2), \quad (-1, -1, -1), \quad (3, 3, 3).$

The corresponding values of l are

$1, \quad 0, \quad 0, \quad 1, \quad 3, \quad 3, \quad 9,$

and the desired result now follows. q.e.d.

We can deduce from lemma 5 that S_2 is connected. We now have the necessary ingredients to give the classification theorem of immersed projective planes of Boy type.

We shall start by defining a particular type of deformation of self-transversal immersions. Let M be a smooth closed surface and $M \xrightarrow{f} \mathbb{R}^3$ a smooth self-transversal immersion. Take two distinct points a, b of the self-intersection set which are not triple points, and two smooth simple paths γ and δ with the following properties:

(i) γ starts at a and ends at b; while δ starts at b and ends at a; both of the paths γ and δ lie on the image of f.

(ii) The two paths arrive at a and b transversally to the self-intersection set and on different sheets of surface.

3.2 Immersed Projective Planes of Boy Type

(iii) The sum $\gamma \cdot \delta$ is a piecewise smooth simple loop bounding a disk D whose interior is C^∞-diffeomorphic to the open disk B^2 and does not meet the image of f.

By an isotopy of $\mathbb{R}^3$, a neighborhood of the disk D can be made to appear as in Fig. 45. Here the image of f is represented by the plane $z = 0$ and the hyperbolic paraboloid $z = (y^2 - x^2)/2 + 1$, while the disk D is given by

$$y = 0 \qquad z \geq 0 \qquad z + x^2/2 - 1 \leq 0.$$

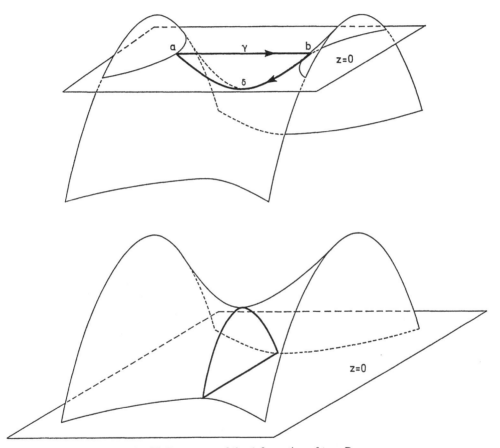

Fig. 45 Two stages of the deformation of type D_1.

The self-intersection set is the hyperbola

$$x^2 - y^2 = 2 \qquad z = 0,$$

and the path γ is the line segment from $(-\sqrt{2}, 0, 0)$ to $(\sqrt{2}, 0, 0)$ while δ is an arc of the parabola

$$y = 0 \qquad z + x^2/2 - 1 = 0.$$

The deformation we want to define is given by a one parameter family f_t of self-transversal immersions of M in $\mathbb{R}^3$, whose images are sent by the above isotopy to the union of the plane $z = 0$ and the hyperbolic paraboloid

$$z = (y^2 - x^2)/2 + t .$$

The equivalence class of f_t changes as t passes through the value zero. This deformation is said to be of *type* D_1 and is one of the six generic deformations of immersed closed surfaces in $\mathbb{R}^3$ [MP].

Theorem 6.

There are precisely two immersed projective planes of Boy type. The first is given by the parametrization F_1 (sec. 2.4) made smooth at the pole; the second is obtained from the first by a deformation of type D_1, followed by a reflexion.

Proof.

By lemma 5, there are only two possibilities to consider

$$(n_1, n_2, n_3) = (0, 0, 0) \quad \text{or} \quad (0, 0, -2) .$$

In each of the two cases, we have fixed the immersion in some neighborhood of pS_2 whose boundary consists of four loops bounding four disjoint disks in $\mathbb{P}^2$. We have to extend the immersion by defining an embedding of each of these disks in S^3.

The complement of the image of $\overline{W}_e$ in S^3 is smoothly diffeomorphic to a *handlebody of dimension* 3 *and genus* 3, i.e. the region lying inside a closed surface of genus three smoothly embedded in $\mathbb{R}^3$. The four loops described above lie on the surface of genus three which bounds the handlebody.

A topological result (see e.g. [BON]) tells us that there is only one isotopy class of embeddings of four disks in such a way that their boundaries are the above loops.

It remains to check that a deformation of type D_1 transforms the self-intersection set of the first immersed projective plane − a direct three-bladed propeller − into an opposite three-bladed propeller; and that after a reflexion, a neighborhood of this propeller satisfies the conditions giving the second immersion of the projective plane [AP]. q.e.d.

We observe that a combinatorial model of the second immersed projective plane of Boy type was constructed in representation 10 of sec. 1.2. Note that in the first immersion the three blades of the propeller bound three embedded disks in the image of the immersion, whereas in the second immersion there are only two such disks. This means in the second case that there cannot be a threefold symmetry, which would no doubt complicate any search for equations similar to that which we carried out for the Boy surface.

Theorem 6 allows us to check that the two model immersed projective planes in $\mathbb{R}^3$ constructed by W. Boy correspond to the same immersion in our classification (see the models in [FI]).

3.3 The Halfway Model

In sec. 2.1 we defined the concept of a regular homotopy between two smooth immersions of a closed surface in $\mathbb{R}^3$. A theorem of Stephen Smale [SM] states that any two smooth immersions of the sphere S^2 in $\mathbb{R}^3$ are regularly homotopic. In particular, there are regular homotopies between the standard embedding of S^2 in $\mathbb{R}^3$ and the antipodal embedding, such a homotopy is called an *eversion of the sphere*.

The existence of eversions of the sphere is not intuitively obvious; for instance, consider the following example.

Example 1.

Any smooth immersion of the circle S^1 in $\mathbb{R}^2$ can be considered as a smooth map $I \xrightarrow{\alpha} \mathbb{R}^2$ such that $\alpha^{(n)}(0) = \alpha^{(n)}(1)$ for all n. The *winding number* of the immersion is defined to be the degree of the map $\alpha'/\|\alpha'\|$ (sec. 1.2). The Whitney-Graustein theorem states that two smooth immersions of S^1 in $\mathbb{R}^2$ are regularly homotopic if and only if they have the same winding number.

The standard embedding of S^1 in $\mathbb{R}^2$ has winding number 1 while that of the antipodal embedding is -1; there is thus no eversion of the circle in $\mathbb{R}^2$.

Several eversions of the sphere are given parametrically in [MO2]. In particular, there is one in which the Boy surface of **Plate 33** occurs as an intermediate step. B. Morin has everted the sphere in a way which is symmetric in time and for which the surface occuring halfway through the eversion has a fourfold axis of symmetry. The immersion corresponding to this halfway stage is described in [MP]; we shall call it the *halfway model*.

The halfway model cannot be a self-transversal immersion since it has a quadruple point; we remark that N. Max and T. Banchoff have proved that any eversion of the sphere must pass through an intermediate stage having a quadruple point [BM].

Using the procedure of sec. 2.4, we are going to give explicit equations for the halfway model. In section 2.4 we constructed a Boy surface of degree six generated by ellipses tangent to a fixed plane and passing through a fixed point called the pole of the surface. The planes of these ellipses generate a cone whose vertex is at the pole; the section of this cone by a plane parallel to the tangent plane at the pole is a tricuspidal hypocycloid.

We recall that each ellipse intersects the plane of the hypocycloid in two points. As we vary the ellipses, the first of these points traces out the circle inscribed in the hypocycloid. The locus of the second point is obtained from the circle as follows.

From each point of the circle, measure out along the tangent to the hypocycloid a distance which is $\pi/2$ out of phase with the distance from the circle to the point of tangency with the hypocycloid (Figs. 31, 33).

The Boy surface is double covered by the sphere and has an axis of threefold symmetry. To get a C^1-immersion of the sphere having an n-fold axis of symmetry, for any n, we shall replace the tricuspidal hypocycloid by a hypocycloid having an appropriate number of cusps.

Drawing inspiration from the case $n = 3$, we define the immersion

$$F_{n,d}(\theta, t) = (1 - td\sqrt{2}\sin n\theta + t^2)^{-1} \cdot \begin{bmatrix} (\sqrt{2}/n)\cos(n-1)\theta + (1 - 1/n) t \cos\theta \\ (\sqrt{2}/n)\sin(n-1)\theta - (1 - 1/n) t \sin\theta \\ 1 \end{bmatrix}$$

This map has the following properties [AP].

Proposition 1.

For $1/\sqrt{n} < |d| < \sqrt{2}$, the mapping $F_{n,d}$ is a C^1-immersion of the sphere in $\mathbb{R}^3$ which is smooth everywhere except at the poles, where it fails to be C^2.

If n is even

The hypocycloid mentioned above has $2n$ cusps. The line $X = Y = 0$ is an axis of symmetry of order $2n$ of the image of $F_{n,d}$, which is a subset of an algebraic surface of degree $4n$. The parametrization $F_{n,d}$ has a multiple point at $(0, 0, (n-1)^2/(n^2 - 2n + 3))$ which has $2n$ preimages given by

$$(\theta, t) = (k\pi/n,\ (-1)^{k+1} \cdot \sqrt{2}/(n-1))\ .$$

If n is odd

The above hypocycloid is a double cover of a hypocycloid with n cusps. The map $F_{n,d}$ is invariant under the antipodal action, and so induces a C^1-immersion of the projective plane $\mathbb{P}^2$ in $\mathbb{R}^3$. The line $X = Y = 0$ is an axis of symmetry of order n of the image of $F_{n,d}$, which is a subset of an algebraic surface of degree $2n$. There is a multiple point on the axis with the coordinates given above, but if we consider the surface as the image of $\mathbb{P}^2$, this point only has multiplicity n.

In addition to the Boy surface (**Plates 39–41**), further examples of these parametrizations are illustrated in **Plate 64 and cover Plate**. If parameter d is equated to zero, we obtain a generalisation of the Roman surface (**Plates 59–63**).

The particular case corresponding to the halfway model is given by $n = 2$ and $d = 1$ (**Plate on cover**). Let us look at it in more detail. Proposition 1 tells us that the image of

$$F_{2,1}(\theta, t) = (1 - \sqrt{2}t \sin 2\theta + t^2)^{-1} \cdot \begin{bmatrix} (1/2)(\sqrt{2} + t)\cos\theta \\ (1/2)(\sqrt{2} - t)\sin\theta \\ 1 \end{bmatrix}$$

lies on an algebraic surface of degree eight.

To determine its equation in homogeneous coordinates $X,\ Y,\ Z,\ T$, we introduce the polynomials

$$A = Z(T-Z) \quad B = 2(X^2 + Y^2) \quad C = 2(Y^2 - X^2) \quad D = 2Z^2 \quad E = 4XY$$

which are invariant under the cyclic group of order four of rotations about the axis.

3.3 The Halfway Model

The five invariants we have just defined satisfy the relation

$$E^2 = B^2 - C^2.$$

Eliminating θ and t we obtain the following irreducible polynomial of degree eight:

$$P(X, Y, Z) = (A-D)^4 + B(-4A^3 + 5A^2B - 4AB^2 + 4B^3) + 2BD(A^2 + 4AD - 3D^2 + \\ + 5B^2 - 11C^2) + D^2(2B^2 + 11C^2) + 2AD(B^2 - 2C^2) + \\ + 2DEC(2B - 11A - D).$$

Note that the intersection of this surface with the tangent plane at the pole $Z = 0$ is given by two quadruple lines passing through the circular points at infinity

$$X + iY = Z = 0 \qquad X - iY = Z = 0.$$

Each of these lines is a cuspidal edge of the surface, with tangent plane $Z = 0$; transverse to this plane and containing this line there is a further imaginary sheet of the surface. Thus the two lines are triple lines of the surface, while the pole is a quadruple point.

As stated in proposition 1, the surface has another quadruple point at $(X, Y, Z, T) = (0, 0, 1, 3)$ whose preimages under $F_{2,1}$ are

$$(k\pi/2, (-1)^{k+1} \cdot \sqrt{2}) \quad \text{where} \quad 0 \leq k \leq 3.$$

Let us look at the apparent contour seen from the pole, leaving aside the astroid and the part lying in the tangent plane at the pole. What remains is the conic projection of the singular curve of the algebraic surface $P = 0$ onto the plane at infinity $T = 0$. This is the algebraic curve of degree eight given by

$$Q(X, Y, Z) = 25B^3(2C + E) + 4CD(2B^2 - 11C^2) + 3ED(16B^2 - 39C^2)$$

Setting

$$X = r \cos \theta \qquad Y = r \sin \theta \qquad Z = 1$$

we get a parametric representation of this curve using polar coordinates

$$r = ((4/5) \sin 2\theta - (3/5) \cos 2\theta) (3 \sin 2\theta + 4 \cos 2\theta)^{1/2} \cdot (2 \cos 2\theta - \sin 2\theta)^{-1/2}.$$

Now, substituting for r in the equation $P = 0$ and writing $u = \tan 2\theta$ we obtain a non-rational parametrization of one quarter of the self-intersection curve of $F_{2,1}$:

$$\begin{bmatrix} X \\ Y \\ Z \\ T \end{bmatrix} = \begin{bmatrix} 5 \cos \theta \cdot (3 - 4u)(1 + u^2)^{1/2}(3u + 4)^{1/2}(2 - u)^{1/2} \\ 5 \sin \theta \cdot (3 - 4u)(1 + u^2)^{1/2}(3u + 4)^{1/2}(2 - u)^{1/2} \\ 25(u - 2)(u^2 + 1) \\ 3u^3 - 288u^2 + 91u - 54 \end{bmatrix}$$

Here $\theta \in [(-1/2) \text{Arctan}(4/3), (1/2) \text{Arctan} 2]$. To get the rest of the curve of self-intersection, make a sequence of quarter turns about the axis of symmetry $0Z$ **(Plate 58)**.

The image of this parametrization is the union of two curves which are of direct three-bladed propeller type (in the sense of sec. 2.2). However, the corresponding algebraic curve is irreducible as is shown by its conic projection $Q = 0$ (Fig. 46). For if this projection were reducible, it would decompose into two quartics, of which one could be ob-

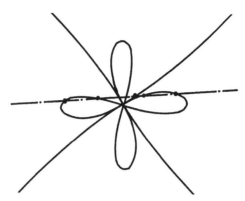

Fig. 46 Conic projection of the self-intersection set of the halfway model. This is the union of two sets of geometric degree 5; a rotation through $\pi/2$ will interchange these two components. The dashed line meets one of these components in five points.

tained from the other by a quarter-turn rotation. But each of the corresponding irreducible components would have *geometric degree* greater than or equal to five; in other words there is a line intersecting the curve transversally at five distinct points.

To conclude, we give some reasons for searching for the simplest possible equations for geometric objects such as the halfway model, the Boy surface, the Klein bottle, or embedded or immersed surfaces in general.

In the last century, such a search was an expression, not always explicitly stated, of a wish to confirm the existence of geometric objects by giving algebraic equations. The algebraic representation was often considered to be the reformulation of a geometric construction in the most rigorous possible terms. This attitude changed with the advent of Analysis Situs.

If an equation is sufficiently simple, it is necessarily conceptual, and in the most desirable cases it will be the answer to a problem which has a unique solution. If this is so, the object under investigation has a representation which can be regarded as canonical. It can then take its place amongst the flora of the geometer.

This flora is more than just a decorative feature in the display cabinets of scientific laboratories. It is also a great source of inspiration for the student as well as for the experienced mathematician. This is one of the great merits of Martin Schilling's catalogue (Katalog mathematischer Modelle für den höheren mathematischen Unterricht, Leipzig 1911). Some admirable photographs of certain of these models appear in [FI].

The computer together with the graphics screen constitute a formidable tool with which we can take a fresh look at the existing set of canonical representations of geometrical objects, and extend the collection by the addition of further figures. However, we must not forget that the use of computer methods can never supplant the mathematical investigation needed in order to find equations which are sufficiently simple. Without such a mathematical investigation, the computer becomes nothing other than the modern day equivalent of plasticene; cleaner, but less easy to use. It is with this point in mind that the plates in this book were created.

Appendix

Listing of the FORTRAN-program used to draw the Boy surface and its deformations

by *Raymond Ripp*

The program was run on a PS 300 Evans-Sutherland connected to a VAX computer.
It generates the Boy surface, the generalizations with an axis of n-fold symmetry, and all the steps of the deformation giving the Roman surface.

Parameters:

G is the parameter of *deformation*
n is the parameter of *symmetry*

$g = 0$	$n = 3$	Plates 21, 22, 56, 57
$g = 1$	$n = 3$	Plates 39, 40, 41, 42, 43
$g = 1/\sqrt{3}$	$n = 3$	Plates 51, 52
$g = 0,4$	$n = 3$	Plate 53
$g = (\sqrt{2}-1)^2$	$n = 3$	Plate 54
$g = 1/1000$	$n = 3$	Plate 55
$g = 1$	$n = 2$	Cover
$g = 0$	$n = 4$	Plates 59, 60, 61
$g = 0$	$n = 5$	Plates 62, 63
$g = 1$	$n = 5$	Plate 64

```
program boy

call parametres

call surface
call intersection

call pssexit
stop
end
```

```
      subroutine parametres

      include 'boycom.inc'

      ne = questionR('$how many ellipses do you want to draw :@')
      np = questionR('$how many points in each ellipse ..... :@')
      g  = questionR('$give the value of g (0 to 1) ........ :@')
      n  = questionI('$give the value of n (n>=2 sym.order) :@')
      npi= questionR('$how many points in the intersection . :@')

      return
      end




      subroutine surface

      include 'boycom.inc'

      logical pl

      pi=3.1415927
      r2=sqrt(2.)

      call newgraph

      do i=0,2*ne-1              ! for each ellipse
          e=float(i)
          pl=.false.
          do j=0,np              ! for each point of the ellipse
              h=float(j)

              a=h*pi/np-pi/2
              b=e*pi/ne

              c=cos(a)/(1-(g*sin(2*a)*sin(n*b))/r2)

              x=c*((r2/n)*cos(a)*cos((n-1)*b)+(n-1)*sin(a)*cos(b)/n)
              y=c*((r2/n)*cos(a)*sin((n-1)*b)-(n-1)*sin(a)*sin(b)/n)
              z=c*cos(a)-2./3.

              call graph(pl,x,y,z)
              pl=.true.
          end do
      end do

      call endgraph

      return
      end
```

FORTRAN-Program for Boy Surface and Deformations

```
subroutine intersection

include 'boycom.inc'

s2=sqrt(2.)

if (n.eq.2)                              call interNeq2

if (n.eq.3.and.g.eq.0)                   call interNeq3Geq

if (n.eq.3.and.g.gt.0.and.g.lt.(s2-1)**2) call interNeq3Glt

if (n.eq.3.and.g.gt.0.and.g.ge.(s2-1)**2) call interNeq3Gge

return
end




subroutine interNeq2    ! intersection N=2

include 'boycom.inc'

logical pl

call newgraph

pl=.false.
do i=0,npi
        ri=float(i)

        t=-0.5*atan(4./3.) + (ri/(2*npi))*(atan(2.)+atan(4./3.))

        c2t=cos(2*t)
        s2t=sin(2*t)

        r=5*(3*c2t-4*s2t)* sqrt( abs((3*s2t+4*c2t)*(2*c2t-s2t)) )

        x0=  3.*s2t*s2t*s2t -288.*s2t*s2t*c2t
1          +91.*s2t*c2t*c2t -  54.*c2t*c2t*c2t
        x1=r*cos(t)
        x2=r*sin(t)
        x3=25*(s2t-2*c2t)

        x=x1/x0
        y=x2/x0
        z=x3/x0-2./3.

        call graph(pl,X,Y,Z)
        pl=.true.
end do
```

```
      pl=.false.
      do i=0,npi
              ri=float(i)

              t=-0.5*atan(4./3.) + (ri/(2*npi))*(atan(2.)+atan(4./3.))

              c2t=cos(2*t)
              s2t=sin(2*t)

              r=5*(3*c2t-4*s2t)* sqrt( abs((3*s2t+4*c2t)*(2*c2t-s2t)) )

              x0=   3.*s2t*s2t*s2t -288.*s2t*s2t*c2t
     1          +91.*s2t*c2t*c2t -  54.*c2t*c2t*c2t
              x1=r*cos(t)
              x2=r*sin(t)
              x3=25*(s2t-2*c2t)

              x=-x2/x0
              y=  x1/x0
              z=  x3/x0-2./3.

              call graph(pl,X,Y,Z)
              pl=.true.
      end do

      pl=.false.
      do i=0,npi




              ri=float(i)

              t=-0.5*atan(4./3.) + (ri/(2*npi))*(atan(2.)+atan(4./3.))

              c2t=cos(2*t)
              s2t=sin(2*t)

              r=5*(3*c2t-4*s2t)* sqrt( abs((3*s2t+4*c2t)*(2*c2t-s2t)) )

              x0=   3.*s2t*s2t*s2t -288.*s2t*s2t*c2t
     1          +91.*s2t*c2t*c2t -  54.*c2t*c2t*c2t
              x1=r*cos(t)
              x2=r*sin(t)
              x3=25*(s2t-2*c2t)

              x=  x2/x0
              y=-x1/x0
              z=  x3/x0-2./3.

              call graph(pl,X,Y,Z)
              pl=.true.
      end do
```

FORTRAN-Program for Boy Surface and Deformations

```fortran
      pl=.false.
      do i=0,npi
            ri=float(i)

            t=-0.5*atan(4./3.) + (ri/(2*npi))*(atan(2.)+atan(4./3.))

            c2t=cos(2*t)
            s2t=sin(2*t)

            r=5*(3*c2t-4*s2t)* sqrt( abs((3*s2t+4*c2t)*(2*c2t-s2t)) )

            x0=  3.*s2t*s2t*s2t -288.*s2t*s2t*c2t
     1          +91.*s2t*c2t*c2t -  54.*c2t*c2t*c2t
            x1=r*cos(t)
            x2=r*sin(t)
            x3=25*(s2t-2*c2t)

            x=-x1/x0
            y=-x2/x0
            z= x3/x0-2./3.

            call graph(pl,X,Y,Z)
            pl=.true.
      end do


      call endgraph

      return
      end



            subroutine interNeq3Geq ! intersection   N=3 G=0

            include 'boycom.inc'

            R2=SQRT(2.)
            R3=SQRT(3.)

            call newgraph
C line 1
            X1=20.*R2/3.
            X2=0.
            X3=-6.-2./3.
            call graph(.false.,X1,X2,X3)

            X1=-20.*R2/3.
            X2=0.
            X3=22./3.-2./3.
            call graph(.true.,X1,X2,X3)

C line 2
            X1=-10.*R2/3.
            X2= 10.*R2/R3
            X3=-6.-2./3.
            call graph(.false.,X1,X2,X3)
```

```
      X1=10*R2/3.
      X2=-10*R2/R3
      X3=22./3.-2./3.
      call graph(.true.,X1,X2,X3)

C line 3
      X1=-10.*R2/3.
      X2=-10.*R2/R3
      X3=-6.-2./3.
      call graph(.false.,X1,X2,X3)

      X1=10.*R2/3.
      X2=10.*R2/R3
      X3=22./3.-2./3.
      call graph(.true.,X1,X2,X3)

      pl=.false.
      do i=1,npi-1
            t=-pi/6. + pi/3.*float(i)/npi
            uscos=1/(3*cos(3*t))
            x=uscos*2*s2*cos(t)
            y=uscos*2*s2*sin(t)
            z=-2./3.
            call graph(pl,x,y,z)
            pl=.true.
      end do

      pl=.false.
      do i=1,npi-1
            t=pi/6. + pi/3.*float(i)/npi
            uscos=1/(3*cos(3*t))
            x=uscos*2*s2*cos(t)
            y=uscos*2*s2*sin(t)
            z=-2./3.




            call graph(pl,x,y,z)
            pl=.true.
      end do

      pl=.false.
      do i=1,npi-1
            t=pi/2. + pi/3.*float(i)/npi
            uscos=1/(3*cos(3*t))
            x=uscos*2*s2*cos(t)
            y=uscos*2*s2*sin(t)
            z=-2./3.
            call graph(pl,x,y,z)
            pl=.true.
      end do

      call endgraph

      return
      end
```

FORTRAN-Program for Boy Surface and Deformations   113

```fortran
        subroutine interNeq3Gge        ! intersection n=3 g>(sqrt(2)-1)**2

        include 'boycom.inc'

        logical pl

        pi=3.1415927
        r2=sqrt(2.)

        call newgraph

        a=atan(g)

        pl=.false.
        do i=0,npi
                t=(pi*float(i))/npi
                apt=a+t

                ss=3*(3*sin(2*a)+sin(6*apt))
                if ( abs(ss).le.0.01 ) then
                                        x=0.
                                        y=0.
                                        z=0.
                                        goto 9
                                end if

                r=(4*r2*cos(a)*sin(3*apt))/ss

                x=r*cos(t)
                y=r*sin(t)
                z=6*sin(2*a)/ss-2./3.

                call graph(pl,x,y,z)
                pl=.true.
        end do

        call endgraph

        return
        end
```

```
      subroutine interNeq3Glt           ! intersection n=3 0<g<(sqrt(2)-1)**:
      include 'boycom.inc'

      logical pl
      real         aa(5000),bb(5000),cc(5000)

      pi=3.1415927
      r2=sqrt(2.)
      r3=sqrt(3.)

      call newgraph

      a=atan(g)

      t1=-asin(3*sin(2*a))/6.-a
      t2= asin(3*sin(2*a))/6.-a+pi/6.
      t3=-asin(3*sin(2*a))/6.-a+pi/3.
      t4=t1
      t5=t2

      do ir=1,2

             do ih=1,npi+1
                   h=float(ih)
                   xh=-1+2*(h)/(npi+2.)
                   t=t4+(t5-t4)*(0.5+xh/(1.+xh*xh))
                   u=3*(3*sin(2*a)+sin(6*(a+t)))
                   v=4*r2*cos(a)*cos(t)*sin(3*(a+t))
                   w=4*r2*cos(a)*sin(t)*sin(3*(a+t))
                   aa(ih)=v/u
                   bb(ih)=w/u
                   cc(ih)=6.*sin(2*a)/u
             end do

             xx1=-20.*r2/3.
             yy1=0.
             xx2=-10.*r2/3.
             yy2=-10.*r2/r3

             do is=0,2
                   pl=.false.
                   if (ir.eq.1.and.g.le.0.01) then
                             x=xx1
                             y=yy1
                             z=22./3.-2./3.
                             call graph(.false.,x,y,z)
                             pl=.true.
                   end if
                   do ih=1,npi
                          x=aa(ih)
                          y=bb(ih)
                          z=cc(ih)-2./3.
                          call graph(pl,x,y,z)
                          pl=.true.
                   end do
                   if (ir.eq.2.and.g.le.0.01) then
                             x=xx2
                             y=yy2
                             z=-6.-2./3.
```

FORTRAN-Program for Boy Surface and Deformations

```
                              call graph(.true.,x,y,z)
                    end if
                if (ir.eq.1.and.g.le.0.01) then
                    x=-(xx1+r3*yy1)/2.
                    y=(-yy1+r3*xx1)/2.
                    xx1=x
                    yy1=y
                    end if
                if (ir.eq.2.and.g.le.0.01) then
                    x=-(xx2+r3*yy2)/2.
                    y=(-yy2+r3*xx2)/2.
                    xx2=x
                    yy2=y
                    end if
                do ih=1,npi+1
                    x=-(aa(ih)+r3*bb(ih))/2.
                    y=(-bb(ih)+r3*aa(ih))/2.
                    aa(ih)=x
                    bb(ih)=y
                    end do
              end do

        t4=t2
        t5=t3
    end do

    call endgraph

    return
    end




subroutine newgraph
include 'amoi:psamoi.inc'
character*2 cnum
        num=num+1
        write(cnum,'(i2.2)')num
        buf='boyobj'//cnum//':=vec item '
        call pssbuf
return
end

subroutine endgraph
include 'amoi:psamoi.inc'
        call pssvsto(0)
return
end

subroutine graph(pl,x,y,z)
logical pl
include 'amoi:psamoi.inc'
        call pssvsto(4,pl,x,y,z)
return
end
```

Bibliography

[AP] Apéry, F.
La surface de Boy, Adv. in Math., Vol 61, No 3, Sept 1986.

[AR] Arnold, V. I.; Gussein-Zade, S. M.; Varchenko, A.
Singularities of differentiable maps, Vol. 1, Monographs in Mathematics, Vol. 82, Birkhauser, 1985.

[BA] Banchoff, T.
Triple points and surgery of immersed surfaces, Proc. Amer. Math. Soc. vol46 n3, p. 407–413, 1974.

[BM] Banchoff, T.; Max, N.
Every sphere eversion has a quadruple point, Contributions to analysis and geometry (Baltimore, Md., 1980), p. 191–209, Johns Hopkins Univ. Press, Baltimore, Md., 1981.

[BON] Bonahon, F.
Cobordism of automorphisms of surfaces, Ann. Sc. Ec. Norm. Sup., 4e série, t. 16, p. 237–270, 1983.

[BOU] Bourbaki, N.
Topologie générale, Hermann, Paris, 1971.

[BOY] Boy, W.
über die Curvatura integra und die Topologie geschlossener Flächen, Math. Ann. 57, 151–184, 1903.

[CE] Cerf, J.
Sur les difféomorphismes de la sphère de dimension trois ($\Gamma_4 = 0$), Lect. Notes in Math. 53, Springer, 1968.

[DA] Darboux, G.
Théorie des surfaces, Gauthiers-Villars, 1914.

[DY] von Dyck, W.
Beiträge zur Analysis situs I, Math. Ann., t. 32, p. 457–512, Leipzig, 1888.

[FI] Fischer, G.
Mathematicals Models, Vieweg, 1986.

[FR] Frégier, M.
Théorèmes nouveaux sur les lignes et surfaces du second ordre, Annales de Gergonne, VI n° VIII fév. 1816.

[GO] Golubitsky, M.; Guillemin, V.
Stable Mappings and Their Singularities, Springer-Verlag New York, 1973.

Bibliography

[GRAM] Gramain, A.
Rapport sur la théorie classique des noeuds (1ère partie), Séminaire Bourbaki, vol. 1975/76, Exposés 471–488), Lectures Notes in Math., 567, Springer, 1977.

[GRAS] Grassmann, H.
Die Ausdehnungslehre vollständig und in strenger Form, Verlag von Th. Enslin, Berlin, 1862.

[GRI] Griffiths, H. B.
Surfaces, Cambridge University Press, 1976.

[HA] Haefliger, A.
Quelques remarques sur les applications différentiables d'une surface dans le plan, Ann. Inst. Fourier, 10, p. 47–60, Grenoble, 1960.

[HI] Hirsch, M. W.
Differential Topology, Springer-Verlag, New York, 1976.

[HO] Hopf, H.
Differential geometry in the large, S. 104, Springer LNM 1000, 1983.

[KL] Klein, F.
Gesammelte Mathematische Abhandlungen, J. Springer, Berlin, 1923.

[MAR] Martinet, J.
Singularities of Smooths Functions and Maps, London Math. Soc., Lect. Note Ser. 58, Cambridge, 1982.

[MAS] Massey, W. S.
Algebraic Topology: An Introduction, Springer-Verlag, New York, 1967.

[MO1] Morin, B.
Formes canoniques des singularités d'une application différentiable, CRAS, t. 260, p. 5662–5665 et 6503–6506, Paris, 1965.

[MO2] Morin, B.
Equations du retournement de la sphère, CRAS série A, t. 287, 879–882, Paris, 1978.

[MP] Morin, P.; Petit, J.-P.
Le retournement de la sphère, Pour la Science, 15, p. 34–49, 1979.

[PE] Petit, J.-P.; Souriau, J.
Une représentation analytique de la surface de Boy, CRAS série I, t. 293, 269–272, 1981.

[PI] Pinkall, U.
Regular homotopy classes of immersed surfaces, to appear in Topology 1986.

[PO] Pont, J.-C.
La topologie algébrique des origines à Poincaré, P.U.F. Paris, 1974.

[RE] Reinhardt, C.
Zu Möbius' Polyedertheorie, Berichte über die Verhandlungen der Königlichen Sächsischen Gesellschaft der Wissenschaften zu Leipzig, Leipzig, 1885.

[SC] Schilling, F.
über die Abbildung der projektiven Ebene auf eine geschlossene sigularitätenfreie Fläche im erreichbaren Gebiet des Raumes, Math. Ann. 92, 69–79, 1924.

[SM] Smale, S.
A classification of immersions of the two-sphere, Transactions A.M.S., 90, p. 281–290, 1959.

[SP] Spivak, M.
A Comprehensive Introduction to Differential Geometry, vol. 1, Publish or Perish, Inc., Berkeley, 1979.

[WA] Wallace Collao, M.
Singularités de codimension deux des surfaces. Thèse de 3e cycle, Publication IRMA, Strasbourg, 1981.

[WE] Weierstrass, K.
Zwei spezielle Flächen vierter Ordnung, Jacob Steiner's Gesammelte Werke Bd. II, S. 741–742.

[WH1] Whitney, H.
On the topology of differentiable manifolds, Lectures in topology, Univ. Mich. press, p. 101–141, Ann Arbor, 1941.

[WH2] Whitney, H.
The General Type of Singularity of a Set of $2n-1$ Smooth Functions of n Variables, Duke Journal of Math., Ser. 2, 45, p. 220–293, 1944.

Subject Index

alternating 60
annulus 2
apparent contour 56
atlas 42

ball (closed unit) 1
ball (open unit) 1
base-point 22
boundary 1
Boy immersion 50
Boy surface (combinatorial) 20
Boy surface (direct) 51, 52, 79
Boy surface (opposite) 51

canonical map 2
class 23
class C^r 43
closed 1
complex 5
0-complex 5
1-complex 5
2-complex 5
confluence (elliptic) 71
confluence (hyperbolic) 71
confluence of Whitney umbrellas 70
connected (pathwise) 22
contingent 63
covering (connected) 28
covering of B of projection p and base B 26
covering (n-sheeted) 27
covering (trivial) 27
critical point 45, 56
cricital set 56
critical value 56
cross-cap 40
cross-cap (combinatorial) 18
curve 1
curve of umbrellas 70, 71

degree 25
degree (geometric) 106
C^r-diffeomorphism 43
differentiable 43
differentiable (r-times) 43
discrete 27
disk 2
disk (open) 2

edge 5
C^r-embedding 44

embedding (topological) 1
equivalent 22, 89
Euler characteristic 5
eversion of the sphere 103
exact 29

fiber above b 26
fiber of the covering 26
flag 14
fold 57
Frégier point 32
fundamental group 23

generic 57
genus 13
germ at a with value in 63
graph 5

half-tangent 91
halfway model 103
handkerchief folded into quarters 56
handle of the Whitney umbrella 63
handlebody of dimension 3 and genus 3 102
Hessian quadratic form 45
homogeneous of even degree 54
homography 14
homotopy 6
homotopy of links 93
homotopy (regular) 50
homotopy type 6
Hopf fibration 84
hypocycloid (elongated) 77

C^r-immersion 48
immersion (topological) 41
index 45
invariant under the antipodal action 54
isotopic (ambient) 41
isotopy (C^r-ambient) 51

Jordan curve theorem 4

Klein bottle 13

lemniscate of Bernoulli 48, 85
lift 57
line 14, 31
line (complex projective) 18
line (projective) 14
link 91

linking number 92
local coordinate system at x 42
loop 9
loop based at 22
loop (constant) 23
loop (simple) 4

manifold (n-dimensional) 1
manifold (without boundary) 1
manifold (smooth) 42
maximum 45
minimum 45
Möbius strip 3, 14
Morse function 45

non-degenerate 45
nondegenerate critical point of index r 61
nonorientable 4

orientable 4, 88
orientation-preserving 89
oval 30

pâquerette de Mélibée 85
path 8
path (opposite) 8
path (piecewise smooth) 91
path (simple) 9
path (smooth) 91
1-pencil of conics 37
2-pencil of conics 36
3-pencil of conics 37
permutation 28
plane (projective) 8
plane (real projective) 8
pleat 57
Plücker conoid 68
pole of the Boy surface 74
pole of the Roman surface 74
projective group 14
projective plane of Boy type (immersed) 90
projective space (n-dimensional) 32
projective space (n-dimenional real) 43

quotient 2
quotient topology 2

rank 44
rank theorem 61
rational 79
resultant 60
retraction 93
Rheinhardt heptahedron 17
Riemannian metric 83
Roman surface 37, 40

saddle 45
saturated 2
segment (unit) 1
self-intersection point 48
self-intersection set 48
self-transversal 84
singular point 62
singularity at a 62
smooth 43
smooth structure on M 42
smooth structure (oriented) 88
smooth vector bundle of dimension 1 with base 94
sphere (unit) 2
stable 57
stable at a 61
starlike 23
Steiner surface 36, 37
strong deformation retract 93
submanifold 86
sum (connected) 11, 50
sum (of paths) 9
sum (topological) 1
support 25, 91
surface 1
surface (embedded) 45
surface (immersed closed) 50
suspended umbrella 70

tacnode 80
tangent bundle 82, 83
tangent bundle (unit) 83
tangent space 82
tangent space to M at a 83
tangent vector to M at a 83
three-bladed propeller 20
three-bladed propeller (direct) 51, 90
tied vector 82
torus 2
transversally 59
trivial 93
tubular neighborhood 93
type D_1 102

umbilic (hyperbolic) 62
unfold 69

Veronese surface 37
Veronese embedding 40
vertex 5

Whitney umbrella 62
Whitehead link 93
winding number 103

Plate Index

1	4, 13, 45	23	63, 68	45	79
2	57	24	68	46	52, 79
3	58	25	72	47	52, 79
4	58	26	72	48	79
5	59	27	72	49	79
6	59	28	72	50	79
7	3, 13, 44	29	72	51	81
8	44	30	72	52	81
9	50	31	55, 74	53	81
10	13, 44, 47	32	55, 74	54	81
11	11, 47	33	103	55	81
12	13, 44, 47	34	55	56	81
13	48	35	55	57	81
14	13	36	60	58	105
15	13, 49	37	60	59	104
16	13, 49	38	60	60	104
17	86	39	78, 104	61	104
18	41, 63	40	78, 79, 81, 104	62	104
19	41, 63	41	78, 79, 81, 104	63	104
20	41, 63	42	78, 90	64	104
21	37, 41, 63, 74	43	78, 90	Cover	104
22	37, 41, 63, 74	44	79		

Plates

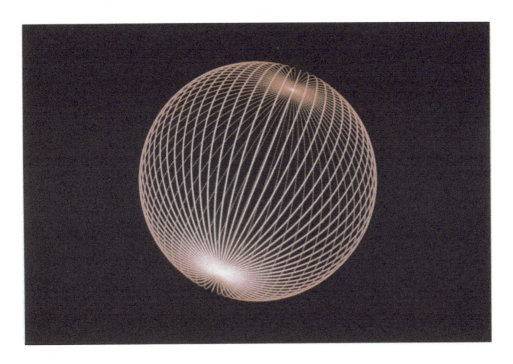

1 Sphere

2 Sphere with three great circles

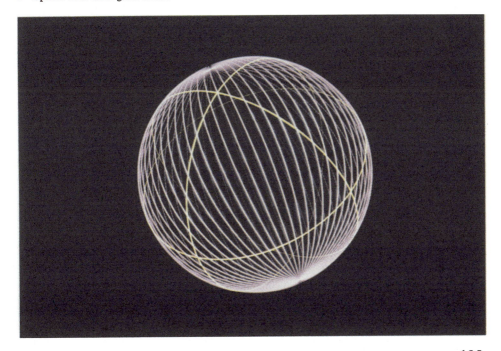

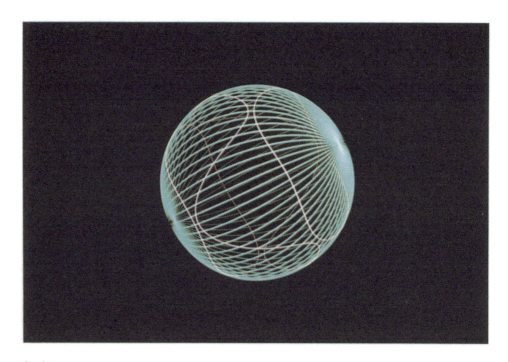

3   Sphere with a nonconnected perturbation of three great circles

4   Nonconnected perturbation of three great circles

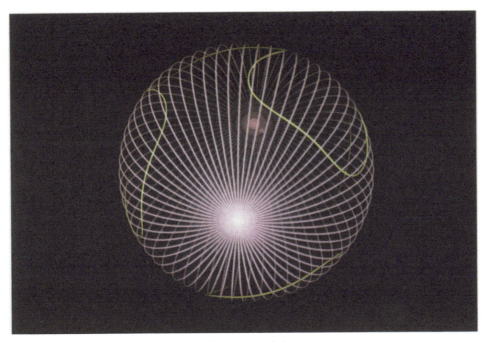

5 Sphere with a connected perturbation of three great circles

6 Connected perturbation of three great circles

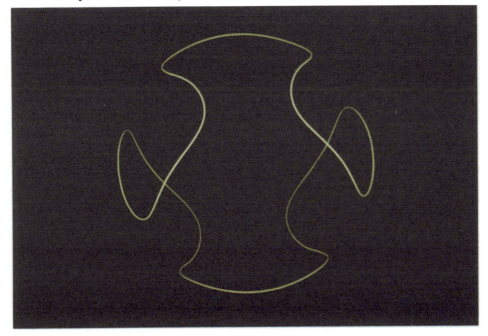

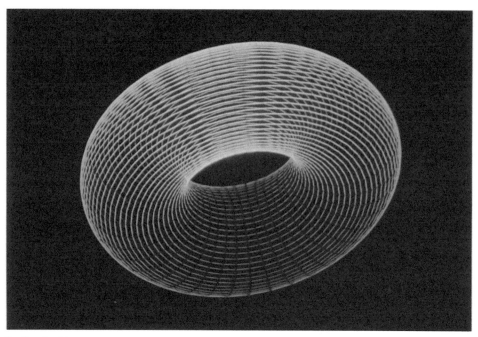

7 Standard torus

8 Torus generated by Villarceau circles

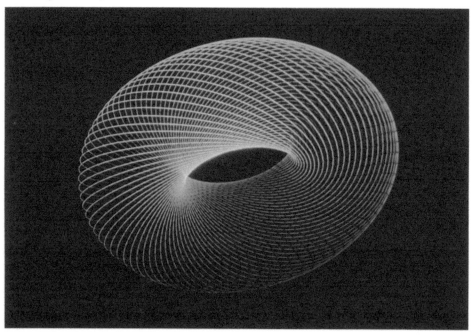

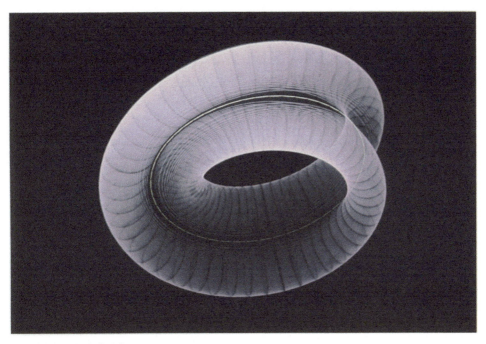

9 Immersed Klein bottle

10 Oriented closed surface of genus two

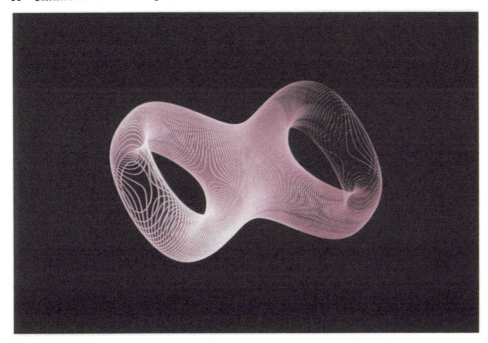

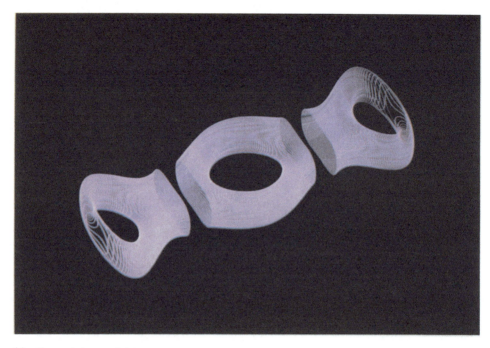

11   Connected sum of three tori

12   Oriented closed surface of genus three

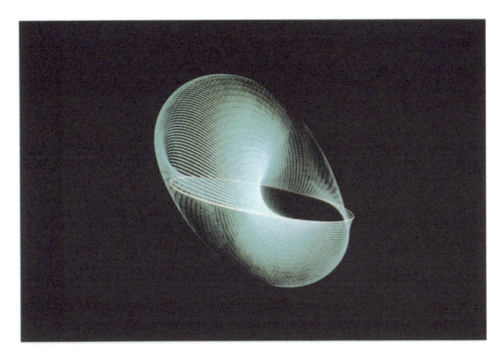

13 Möbius strip with a circle as boundary

14 Immersed Möbius strip

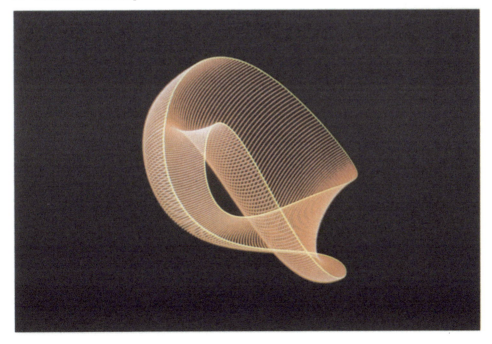

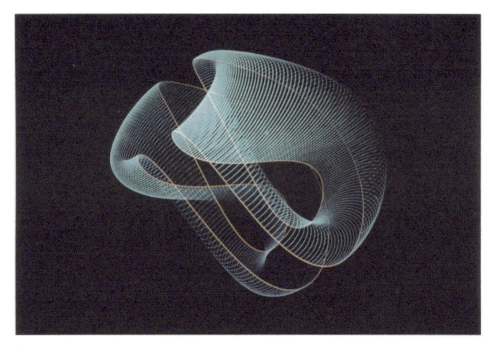

15 Gluing two Möbius strips along their boundaries

16 Klein bottle

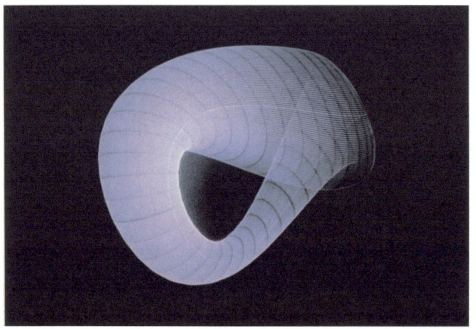

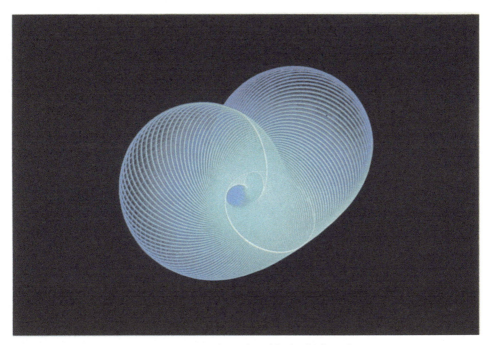

17 Klein bottle obtained by gluing two Möbius strips with circular boundary

18 Steiner cross-cap

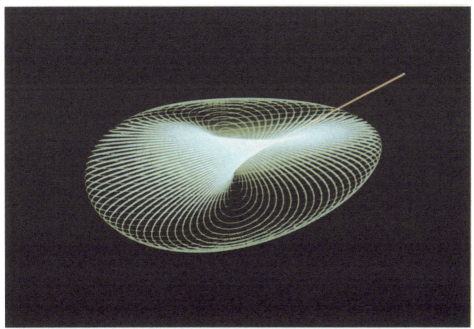

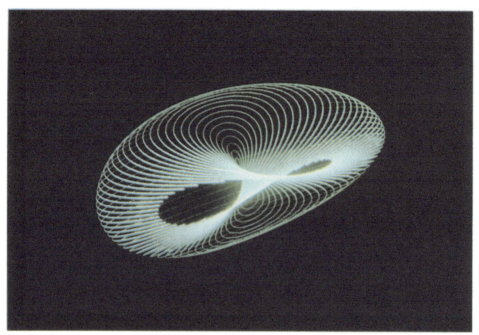

19 Steiner cross-cap with a window

20 Steiner cross-cap

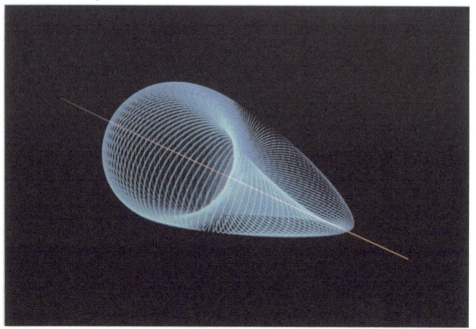

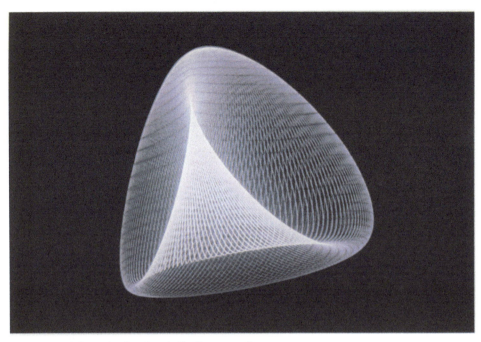

21 Image of the projective plane in the Roman surface

22 Roman surface

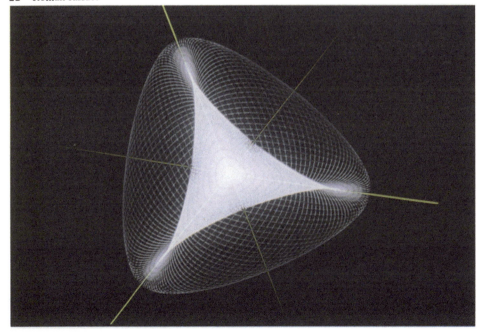

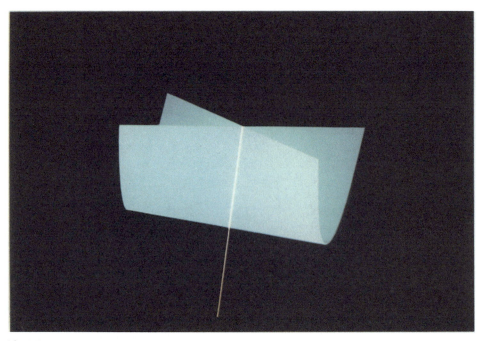

23 Whitney umbrella on the ruled cubic surface

24 Plücker conoid

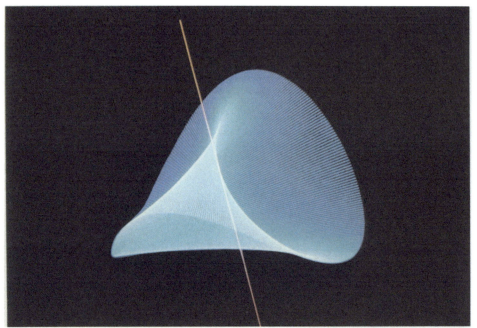

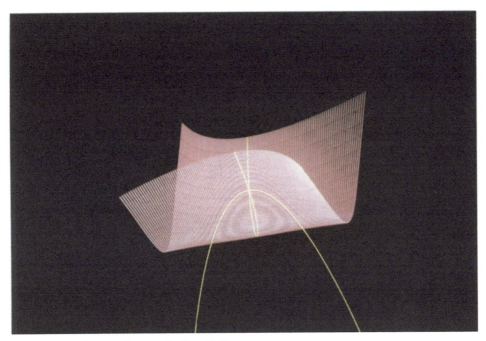

25 Elliptic confluence of two umbrellas: $t > 0$

26 Elliptic confluence of two umbrellas: $t = 0$

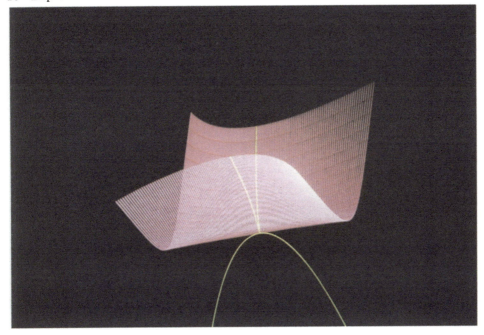

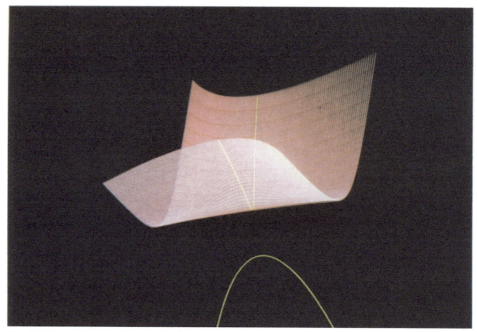

27 Elliptic confluence of two umbrellas: t < 0

28 Hyperbolic confluence of two umbrellas: t > 0

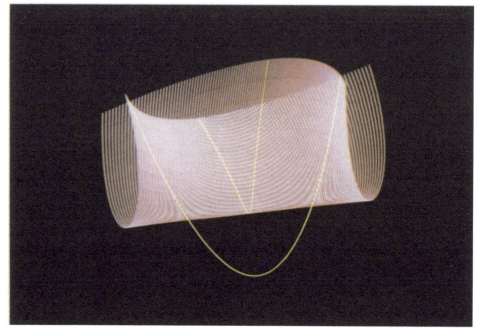

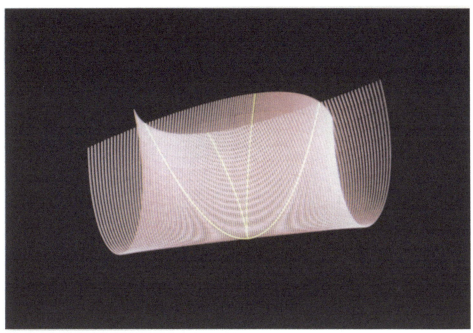

29 Hyperbolic confluence of two umbrellas: t = 0

30 Hyperbolic confluence of two umbrellas: t < 0

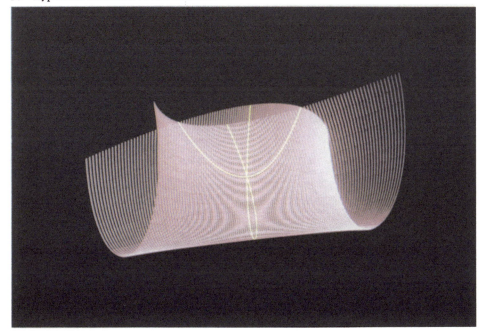

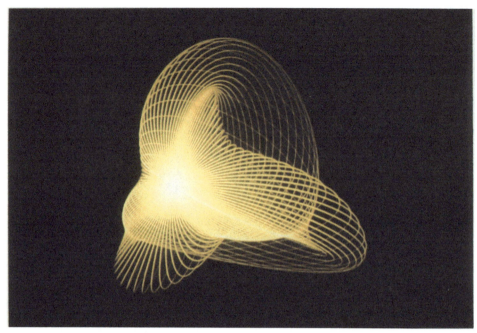

31 Boy surface according to Petit/Souriau [PE]

32 Boy surface according to Petit/Souriau [PE]

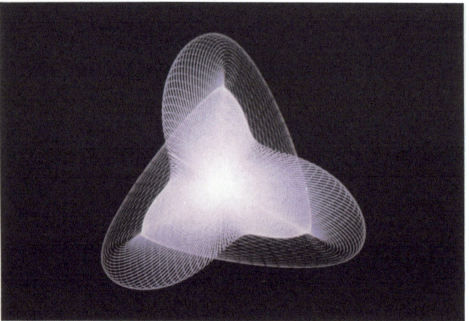

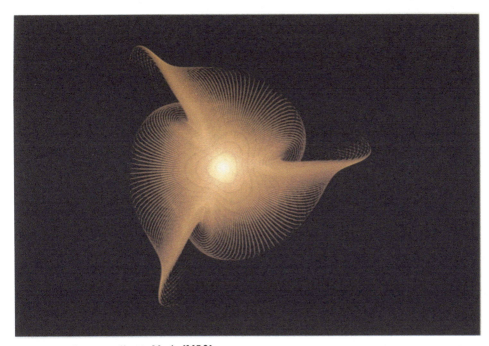

33 Boy surface according to Morin [MO2]

34 Boy surface according to a parametrization due to J. F. Hughes

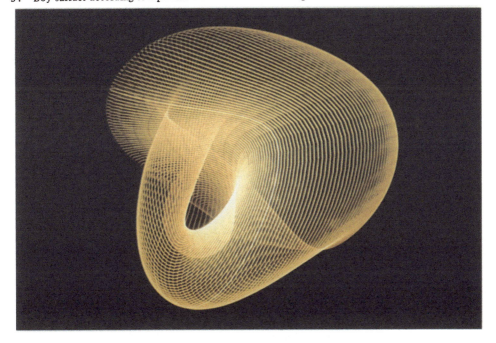

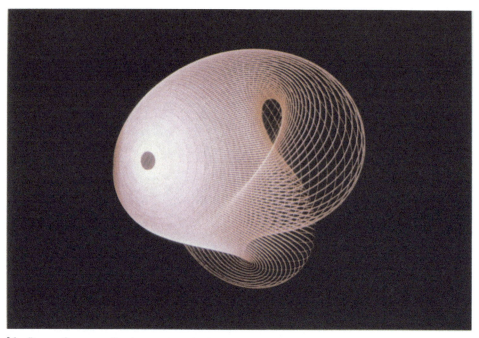

35 Boy surface according to a parametrization due to R. Bryant

36 Boy surface parametrized by three homogeneous polynomials of degree four on the sphere

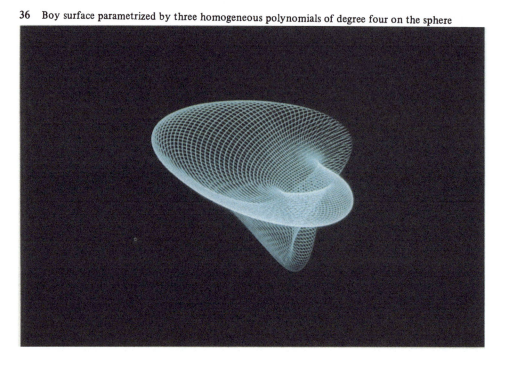

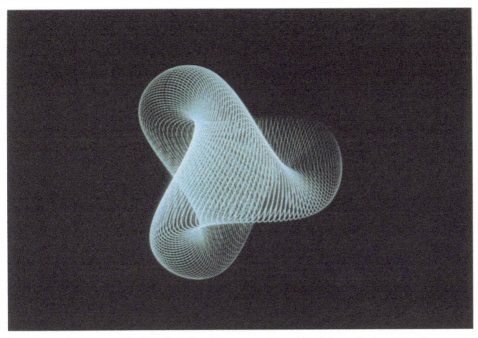

37 Boy surface parametrized by three homogeneous polynomials of degree four on the sphere

38 Boy surface parametrized by three homogeneous polynomials of degree four on the sphere with a window

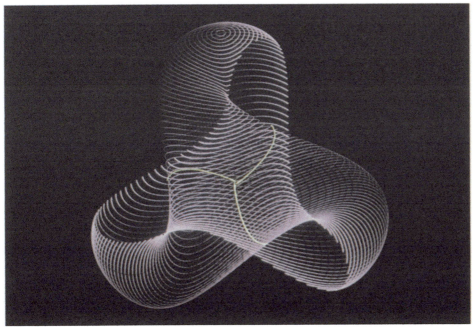

143

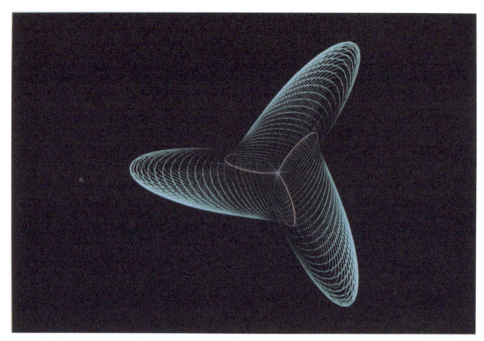

39 Boy surface of degree six

40 Boy surface of degree six

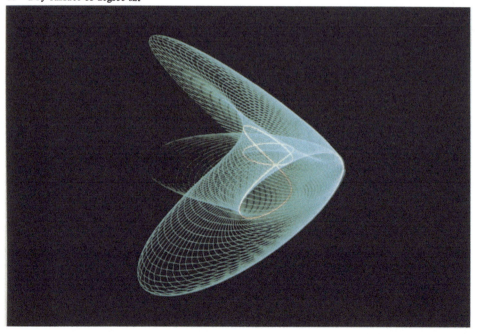

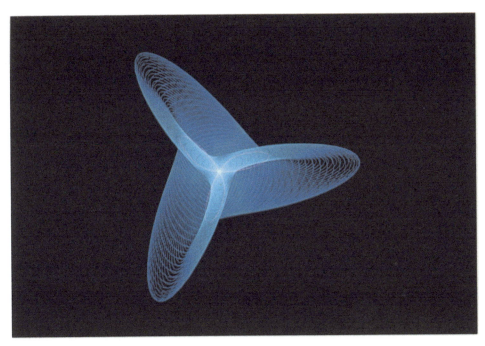

41 Boy surface of degree six

42 Curves of construction of the Boy surface of degree six

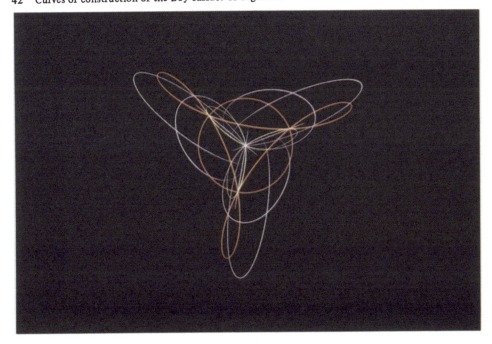

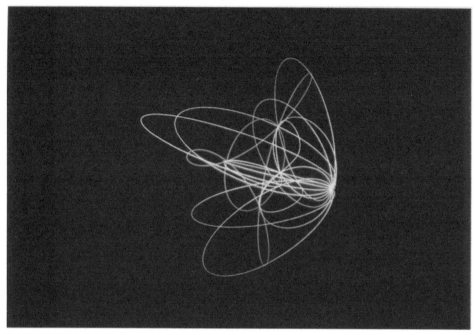

43 Curves of construction of the Boy surface of degree six

44 Boundary component of a neighborhood of the self-intersection set of the Boy surface

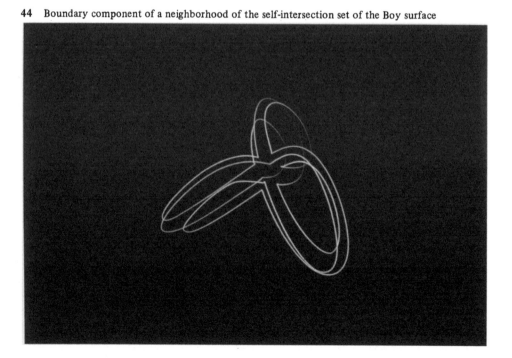

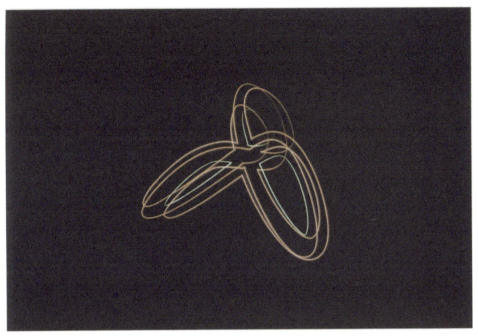

45 Boundary of a neighborhood of the self-intersection set of the Boy surface

46 Neighborhood of the self-intersection set of the Boy surface

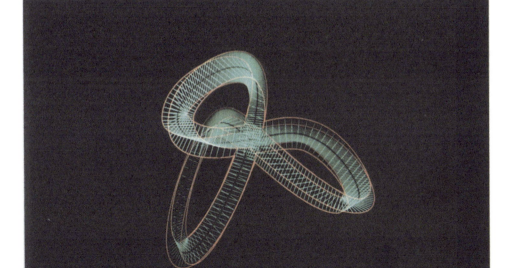

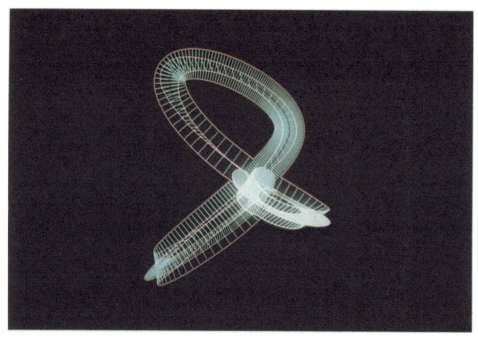

47 Neighborhood of the self-intersection set of the Boy surface

48 Level curves of the Boy surface situated between the plane of saddles and the pole

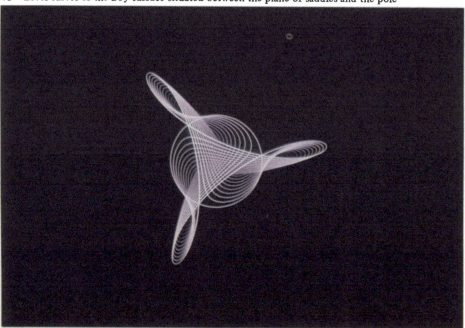

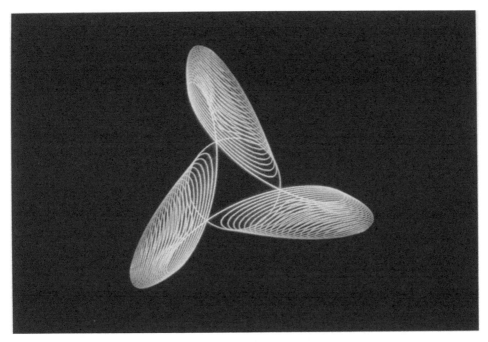

49 Level curves of the Boy surface situated between the plane of saddles and the plane of minima

50 Level curves of the Boy surface

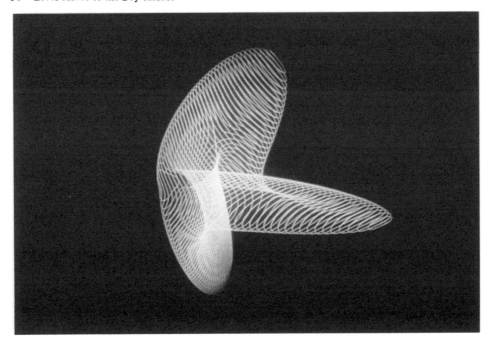

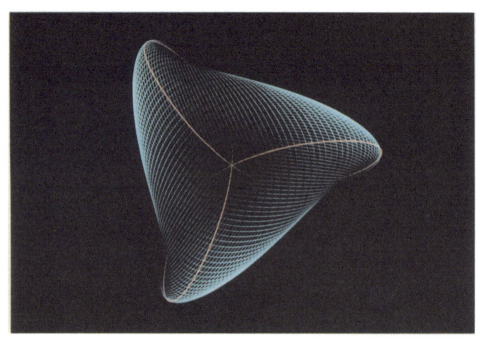

51 Confluence of umbrellas on the Roman surface $d = 1/\sqrt{3}$

52 Confluence of umbrellas on the Roman surface $d = 1/\sqrt{3}$

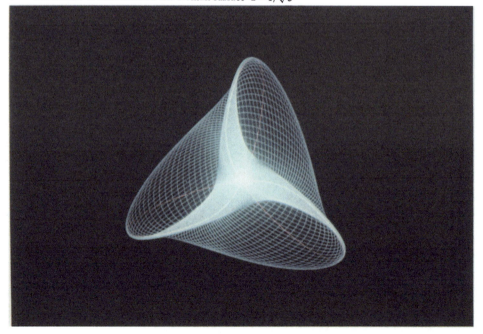

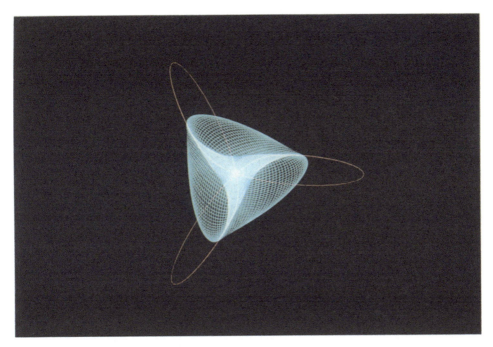

53 Deformation of the Roman surface d = 0.4

54 Deformation of the Roman surface: the self-intersection curve is tangent to the plane at infinity $d = (\sqrt{2} - 1)^2$

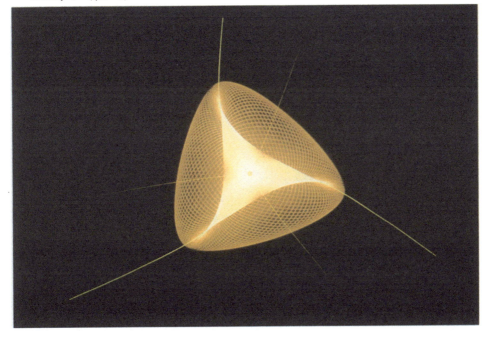

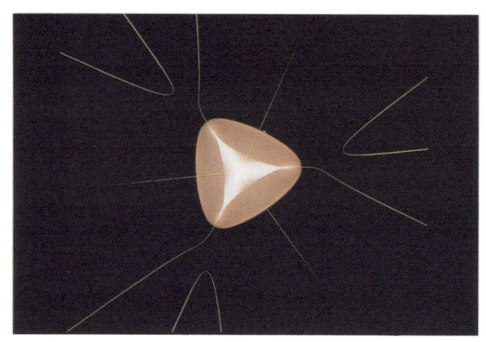

55 Deformation of the Roman surface d = 0.001

56 Beginning of the deformation of the Roman surface d = 0

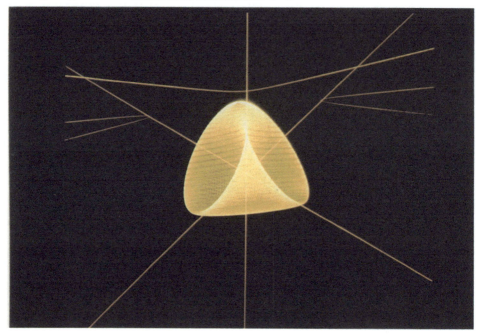

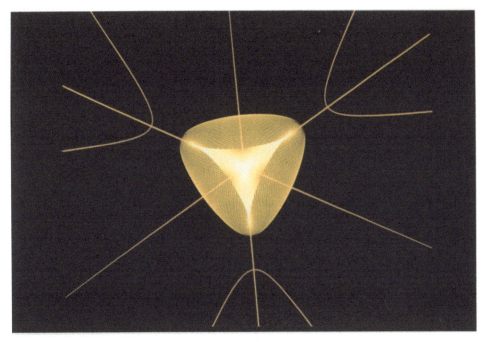

57 Beginning of the deformation of the Roman surface d = 0

58 Self-intersection set of the halfway model

59 Surface of Roman type having an eightfold symmetry

60 Surface of Roman type having an eightfold symmetry

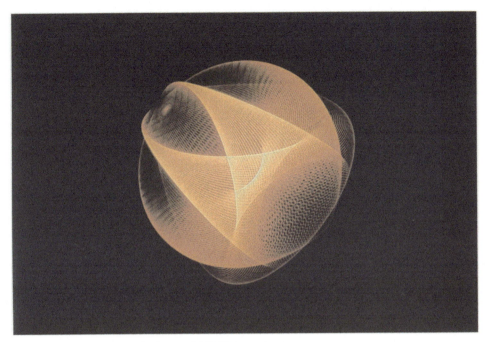

61 Surface of Roman type having an eightfold symmetry

62 Surface of Roman type having a fivefold symmetry with window

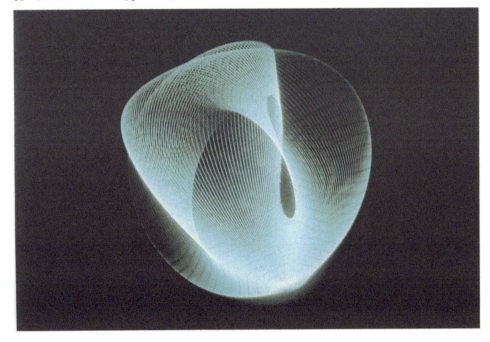

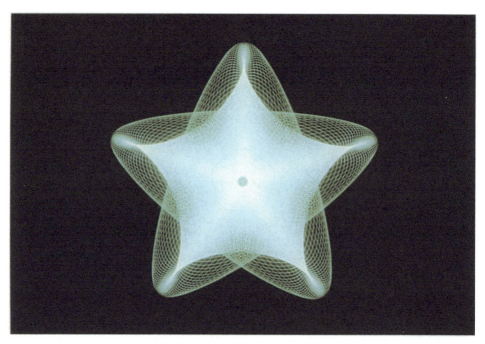

63   Surface of Roman type having a fivefold symmetry

64   Immersed projective plane having a fivefold symmetry

Printed by Printforce, the Netherlands